Homogeneous Ordered Graphs, Metrically Homogeneous Graphs, and Beyond

Volume II: 3-Multi-graphs and 2-Multi-tournaments

This is the second of two volumes by Professor Cherlin presenting the state of the art in the classification of homogeneous structures in binary languages and related problems in the intersection of model theory and combinatorics. Researchers and graduate students in the area will find in these volumes many far-reaching results and interesting new research directions to pursue.

This volume continues the analysis of the first volume to 3-multi-graphs and 3-multi-tournaments, expansions of graphs and tournaments by the addition of a further binary relation. The opening chapter provides an overview of the volume, outlining the relevant results and conjectures. The author applies and extends the results of Volume I to obtain a detailed catalogue of such structures and a second classification conjecture. The book ends with an appendix exploring recent advances and open problems in the theory of homogeneous structures and related subjects.

GREGORY CHERLIN is Distinguished Professor Emeritus at Rutgers University. He has worked on applications of model theory to algebra and combinatorics for half a century, and has published four books and over 100 articles on model theory and its applications.

LECTURE NOTES IN LOGIC

A Publication of The Association for Symbolic Logic

This series serves researchers, teachers, and students in the field of symbolic logic, broadly interpreted. The aim of the series is to bring publications to the logic community with the least possible delay and to provide rapid dissemination of the latest research. Scientific quality is the overriding criterion by which submissions are evaluated.

More information, including a list of the books in the series, can be found at http://aslonline.org/books/lecture-notes-in-logic/

LECTURE NOTES IN LOGIC 54

Homogeneous Ordered Graphs, Metrically Homogeneous Graphs, and Beyond

Volume II: 3-Multi-graphs and 2-Multi-tournaments

GREGORY CHERLIN

Rutgers University, New Jersey

ASSOCIATION FOR SYMBOLIC LOGIC

CAMBRIDGE
UNIVERSITY PRESS

CAMBRIDGE
UNIVERSITY PRESS

University Printing House, Cambridge CB2 8BS, United Kingdom

One Liberty Plaza, 20th Floor, New York, NY 10006, USA

477 Williamstown Road, Port Melbourne, VIC 3207, Australia

314–321, 3rd Floor, Plot 3, Splendor Forum, Jasola District Centre,
New Delhi – 110025, India

103 Penang Road, #05–06/07, Visioncrest Commercial, Singapore 238467

Cambridge University Press is part of the University of Cambridge.

It furthers the University's mission by disseminating knowledge in the pursuit of
education, learning, and research at the highest international levels
of excellence.

www.cambridge.org
Information on this title: www.cambridge.org/9781009229487
DOI: 10.1017/9781009229500

Association for Symbolic Logic
Richard A. Shore, Publisher
Department of Mathematics, Cornell University, Ithaca, NY 14853
http://aslonline.org

First published 2022

A catalogue record for this publication is available from the British Library.

ISBN – 2 Volume Set 978-1-009-23018-6 Hardback
ISBN – Volume I 978-1-009-22969-2 Hardback
ISBN – Volume II 978-1-009-22948-7 Hardback

In memory of GEORGE CHERLIN, 1924–1992.
Enthralled by the beauty of mathematics,
ever mindful of its power for good or evil.

CONTENTS

Volume I

Volume II

LIST OF TABLES

ABSTRACT

Volume I. Part I: A complete classification of homogeneous ordered graphs is given: up to a change of language each is either a generically ordered homogeneous graph or tournament, or a generic linear extension of a homogeneous partial order.

Part II: A catalog of the currently known metrically homogeneous graphs is given, with proofs of existence and some evidence for the completeness of the catalog. This includes a reduction of the problem to what may be considered the generic case, and some tools for the analysis of the generic case.

Some related developments are discussed in an appendix.

Volume II. Here the impact of the results of Parts I and II and of related work in Amato, Cherlin, and Macpherson [2021] on the classification of homogeneous structures for a language with two anti-symmetric 2-types or with 3 symmetric 2-types is worked out in detail.

An appendix to Volume II discusses some further advances in related areas, and a wide variety of open problems.

An extensive bibliography of related literature and a quick survey of that literature, organized by topic, is given in Cherlin [2021] (see also http://www.cambridge.org/9781009229692).

The method used in Part I of Volume I is due to Alistair Lachlan. The method used in Part II of Volume I and throughout Volume II is a direct application of Fraïssé's theory of amalgamation classes.

2020 Mathematics subject classifications. Primary 03C15. Secondary 03C10, 03C13, 05C12, 05C55, 05C63, 06F99, 20B27.

PREFACE TO VOLUME II

A comprehensive introduction to both volumes of this work is given in Volume I. We give a briefer introduction to Volume II here.

In Volume I we considered two classification problems, the classification of the homogeneous ordered graphs and the classification of metrically homogeneous graphs, in Parts I and II respectively. We gave a full solution to the first problem and provided an explicit catalog and conjecture, a body of relevant theory, and some applications in Part II. Additional work in Amato, Cherlin, and Macpherson [2021] confirms the conjecture in diameter 3 and further discussion of the prospects for the general case is found in the appendix to Volume I.

As we explained in the preface to Volume I, the work in that volume together with the work in diameter 3 casts some light on the classification of the homogeneous structures with two pairs of anti-symmetric 2-types or with 3 symmetric 2-types, which we call 2-*multi-tournaments* or 3-*multi-graphs*, respectively. The task of the present volume is to see where exactly these results leave us, with respect to the broader problems. Clarifying this point requires us to undertake some further substantial explorations, some of which might be considered logically prior to the material of Volume I, from the point of view of a systematic study.

Recall that we work throughout with amalgamation classes in accordance with the Fraïssé theory, which amounts to characterizing homogeneous structures by their forbidden substructures. We call a homogeneous structure 3-*constrained* if the minimal forbidden substructures have order at most 3. In dealing with explicit classification problems for binary relational languages, leaving aside the specialized techniques useful for some particular classes (such as finite structures), the general approach to making a catalog (potential classification) of homogeneous structures and investigating its completeness is the following.

1. Classify the 3-constrained homogeneous structures of the desired type.

2. Show that, with few exceptions, the triangle constraints in any homogenous structure of the desired type agree with the triangle constraints in some 3-constrained structure.

3. In all cases for which the triangle constraints are inconsistent with free amalgamation, classify the resulting structures, which are considered to be of exceptional type from the point of view of the general problem.

4. In the remaining case, show, again with few exceptions, that the remaining homogeneous structures are associated with free amalgamation classes.

If this last step ever breaks down—as well it might—we should see something distinctly new appearing. However our focus now is on the first three steps.

The first step was actually carried out in Cherlin [1998], with computer assistance. In the case of 2-multi-tournaments we have to redo this by an explicit argument in order to prepare the way properly for step (2).

One notices that the work carried out in Part I of Volume I and in Amato, Cherlin, and Macpherson [2021] bears specifically on instances of step (3) in each of the two contexts, and that there are indeed some other points that have to be addressed in order to draw definite conclusions as to where the broader classification stands.

This explains why we felt the present volume was necessary. We had thought it would be a supplementary third part to the two parts of Volume I, but discovered that there was a great deal of unfinished business still to be taken care of to bring this to a satisfactory state.

What was achieved, and how, will be described in considerable detail in Chapter 18 (we continue the numbering of chapters, and of open problems, from Volume I). Here we give an overview.

The results for homogeneous 3-multi-graphs are quite satisfactory. They occupy Chapter 19, and are summarized in §19.4. We find that an unknown homogeneous 3-multigraph must be infinite and primitive, and the only forbidden triangles are monochromatic (mainly due to the classification theorem of Amato, Cherlin, and Macpherson [2021], as well as prior classifications in finite and imprimitive cases). These constraints must also be compatible with free amalgamation. With this as a point of departure one may expect, or in any case hope, to find that what remains are in fact free amalgamation classes. At that stage, the powerful classical methods of Lachlan and Woodrow may well become relevant.

On the other hand, in the case of 2-multi-tournaments, we find we have a good deal more work to do. At the end of our analysis four very recalcitrant cases remain, which do not correspond to 3-constrained examples and therefore should, conjecturally, be eliminated.

To begin with, there is no prior classification of homogeneous 2-multi-tournaments in the imprimitive case, so we supply one in Chapter 20. Then we give a proof of the classification of the 3-constrained homogeneous 2-multi-tournaments (Chapter 21). This is needed because we require more than the resulting list of examples—we need to understand why there are no other examples under the strong assumption of 3-constraint. The next challenge is to reach the same conclusion concerning triangle constraints under the assumption of homogeneity alone.

Here a lengthy analysis of the possible patterns of forbidden triangles leads to the identification of four potentially exceptional cases, not corresponding to 3-constrained structures, but not yet ruled out. See Proposition 22.1.1 and §22.8, as well as §22.9.1.

If one uses explicit amalgamation arguments as we do here, one finds that the exceptional cases are the ones for which every amalgamation diagram on five points can be completed, but not every such diagram on six points can be. In all other cases one can use a series of amalgamation diagrams of order at most five to carry out the analysis; these have factors of order four which while compatible with the triangle constraints may or may not embed in the given structure, but in all cases one arrives eventually at a contradiction.

Leaving aside these four delicate cases—that is, moving on to step (3) of our plan—and returning to the patterns of forbidden triangles which *do correspond* to a 3-constrained class without free amalgamation, and which are *not already covered by Part I*: these are the seven examples shown in Table 21.1, under Groups III and IV, labeled 6–12 (as part of a longer list given earlier).

Now Proposition 22.4.1 gives the classification of the homogeneous 2-multi-tournaments falling under case #12 in Table 21.1. So in order to reach what one may consider the generic case of the classification problem for homogeneous 2-multi-tournaments, one must not only eliminate the four exotic patterns of forbidden triangles already identified, but complete six additional classification problems defined by patterns of forbidden triangles which correspond to 3-constrained homogeneous 2-multi-tournaments. We have not assessed the complexity of these problems. The ones treated in Part I of Volume I were complex, the one just mentioned as case #12 is relatively straightforward, and the rest deserve further study.

Princeton, August 2021

ACKNOWLEDGMENTS

I am grateful to Lionel Nguyen Van Thé for raising the question treated in Part I of Volume I in summer 2012. This did not strike me immediately as a reasonable question, but my inability to articulate a concrete objection soon forced me to take it seriously. I am also grateful to Miodag Sokić for drawing my attention to the article by Dolinka and Mašulović [2012] in Fall 2013.

With regard to the material of Part II of Volume I, I first encountered the problem of classifying countable metrically homogeneous graphs in Moss [1992], then found the discussion in Cameron [1998] very stimulating, and found evidence in Kechris, Pestov, and Todorcevic [2005] that the problem was natural from several other points of view.

Much of my fascination with the classification of homogeneous structures can be traced back to Alistair Lachlan. I have long enjoyed his understated enthusiasm, his gift for turning examples into theories, and his fearlessness.

In the course of preparing this text, remarks by Dugald Macpherson on the content and Jan Hubička on the literature were very helpful.

Lately Braunfeld, Coulson, Evans, Hubička, Konečný, Nešetřil, and Simon have been among those who have complicated my life by telling me interesting things outside the scope of the monograph that nonetheless deserved mention here and there. A good deal of that found its way into the appendix to Volume I. That list could be longer, but one has to stop somewhere.

I greatly appreciate the careful work done by a number of anonymous referees, as should the reader. (This applies with particular force to Volume II.)

For that matter, I also appreciate the work done by an anonymous referee on Amato, Cherlin, and Macpherson [2021] and the impetus provided to us jointly to reflect further on the path forward with regard to the classification of metrically homogeneous graphs.

I thank Stewart Cherlin for editing the family photographs used as a frontispiece to be suitable for reproduction in black and white.

And I take this opportunity to acknowledge, for once, my lifelong debt to Donald Knuth for the years he took off to make TeX work.

Amélie and Grégoire provided regular and varied distractions; le petit Nicolas came in at the end with music and dance. S. L. Huang was a source of additional entertainment. Christiane and Rufus provided their famous hospitality, and Chantal kept it all together.

Research supported by the Hausdorff Institute for Mathematics, Bonn, Fall 2013 and again in Fall 2018, and by a series of grants from the NSF, including DMS-1362974.

Chapter 18

CLASSIFICATION PROBLEMS FOR SMALL BINARY LANGUAGES

In this volume we will consider the material of Parts I and II, together with the result of Amato, Cherlin, and Macpherson [2021], in a substantially broader context, which had been touched on in the Appendix to Cherlin [1998]: namely, lists of 3-constrained structures for small languages were given there, leaving aside the straightforward cases of imprimitive structures or structures with free amalgamation.

I had assumed at first that the consideration of these classification problems would be one of many open problems to be passed in review, in the manner of the appendix to Volume I or the present volume. It quickly became clear that due diligence requires a good deal more in these cases. So clearing up this point, as far as I was able, has become the subject of this volume.

The present Chapter provides an overview of the volume, stating the relevant facts and conjectures, and some associated concrete open questions. The following chapters then prove the various points that have not been dealt with previously.

It is reasonable to ask whether the class of homogeneous structures for a finite relational language is classifiable in any sense at all in general: a reasonable interpretation of this question was given by Lachlan, and we review his suggestion in §18.1, along with another related problem.

Descending (abruptly) to the level of structures homogeneous for a binary relational language, one may formulate the question more narrowly, asking in effect whether there are any examples fundamentally different from those we have already seen. We take a stab at formulating this question precisely in §18.2. In the symmetric case (that is, all 2-types are symmetric) one can go further, as discussed already in the appendix fo Volume I.

After these general considerations we return to the main line of the present monograph. From the point of view of Cherlin [1998], Volume I of the present monograph deals with particular fragments of natural classification problems of a type similar to the one treated in that memoir. Namely, after the classification of homogeneous directed graphs, the next two classification problems to consider in the same spirit would be either the classification of

1

2-multi-tournaments or 3-multi-graphs. By this we mean tournaments with a coloring of the arcs by two colors, or complete graphs with a coloring of the edges by 3 colors, respectively. In model theoretic terms, we would refer to the class of structures with two pairs of anti-symmetric atomic 2-types, or with three symmetric 2-types, respectively. Our choice of terminology here is intended to avoid major clashes with a number of similarly named notions found in the literature. In particular, the term 3-*graph* has been used in the homogeneous context both for graphs with edge colorings and for 3-hypergraphs, while 2-*graph* conventionally has yet another meaning.

As we will discuss, the material of Part I falls within the classification problem for 2-multi-tournaments, and the material of Part II, when one sets $\delta = 3$, falls within the classification problem for 3-multi-graphs; that special case is resolved in Amato, Cherlin, and Macpherson [2021]. Therefore it is natural to take up those two broader problems here, to take stock of the situation, and to continue the analysis somewhat further in each case.

The first step in any classification problem is to make a catalog, or, as Cameron puts it, a "census" of the known examples, which includes making known any that ought to be known at the outset. From the point of view of the classification problems for homogeneous 2-multi-tournaments or 2-multi-graphs, everything in Part I, or in Part II with $\delta = 3$, falls under the heading of census-taking, without however exhausting the subject. Before leaving the subject, we bring the census to a natural stopping point (but not completion). This is discussed in §18.3, and then the work is carried out in Chapter 19 for 3-multi-graphs and in Chapters 20–22 for 2-multi-graphs.

There are many striking conjectures on homogeneous structures for a finite relational language. We have little immediate prospect of settling any of them. But the results of classification projects have provided some examples which have turned out to be helpful in stimulating the development of the relevant techniques.

Notably, since the completion of the first draft of this monograph, a number of lines of development have contributed to a better understanding of the known binary homogeneous structures. These have been discussed in the Appendix to Volume I but we recall them here.

The first line is the application of the theory of ample generics and structural Ramsey theory via a close study of the "partial" metrically homogeneous graphs. This involves work of Aranda Lopez, Bradley-Williams, Coulson, Evans, Hng, Hubička, Karamanlis, Kompatscher, Konečný, Nešetřil, and Pawliuk. A second line is the study of stationary independence relations in the sense of Tent and Ziegler, and involves work of Li.

These two lines, and a good deal of other prior work, lead to a very broad concept of generalized metric space with values in a finite partially ordered commutative semigroup which can account for the "exotic" 3-constrained examples in the appendix to Cherlin [1998] and also helps in understanding the properties of the 3-constrained metrically homogeneous graphs of generic type. This setting includes the generalized ultrametric spaces used by Braunfeld in studying Cameron's problem on structures with finitely many linear orders, and Conant's generalized metric spaces. More may be found in the Appendix to Volume I.

This development reinforces the impression that to date, the stock of examples one has is not all that varied. It remains to be seen whether this way of viewing these examples is limited to the case of symmetric binary languages or can be extended to cover anti-symmetric cases as well (but here one could use additional examples as test cases).

One should also take note of a radically different approach due to Pierre Simon which uses ideas of geometric neostability theory. While not yet very general, it yields results that seem inaccessible by more direct methods when it applies.

18.1. Lachlan's classification problem

Lachlan posed "the" classification problem for homogeneous structures in finite relational languages in its broadest formulation.

PROBLEM 9 (Lachlan). *Is the following problem decidable?*

Given two finite collections of structures A_+, A_- in a finite relational language, determine whether there is an amalgamation class containing A_+ and disjoint from A_-.

In other words, is the classification of homogeneous structures in finite relational languages an art or a science?

One path to decidability might be as follows: one can check easily whether a particular amalgamation problem has a solution. If one could bound the sizes of the amalgamation problems that need to be checked, this would suffice.

For a specified relational language, one may dispose of the problem by any sort of reasonably explicit full classification. For example, there are uncountably many homogeneous directed graphs, but the explicit classification given in Cherlin [1998] passes Lachlan's test: it immediately gives a decision procedure for Lachlan's problem in that setting.

A variant of Lachlan's problem arises as a practical matter in the course of explicit classifications.

PROBLEM 10. *Is the following problem decidable?*

Given a finite collection of structures A_- in a finite relational language, determine whether the class of finite structures containing no isomorphic copy of a structure in A_- is an amalgamation class.

In binary languages this problem is decidable, as it suffices to consider amalgamation problems with just two points outside the base, and one can list the possible obstructions to completion in this case. This is helpful: several explicit amalgamation arguments in each of the first two parts of the present monograph were found by following this line of analysis.[1]

Other variants of Lachlan's problem are of combinatorial interest, notably the following.

PROBLEM 11. *Given $\mathcal{A}_+$, $\mathcal{A}_-$ as above, determine whether the number of amalgamation classes containing $\mathcal{A}_+$ and disjoint from $\mathcal{A}_-$ is*
    (*a*) *finite;*
    (*b*) *countable.*

The decidabilty of version (*b*) of the problem is open even in the case of homogeneous directed graphs, where it becomes equivalent to an instance of the following (for the class of tournaments).

PROBLEM 12 (WQO Problem). *Let $\mathcal{A}_-$ be a finite set of finite structures in a fixed finite relational language, and let $\mathcal{Q}$ be the quasi-ordered class of finite structures not containing any structure in $\mathcal{A}_-$, ordered by isomorphic embedding. Determine whether the class $\mathcal{Q}$ contains an infinite antichain (equivalently, whether the class $\mathcal{Q}$ is "well quasi-ordered"—wqo).*

I have discussed this problem at length in Cherlin [2011a], so I will not elaborate here. It is also well known in the context of permutation pattern classes, where the wqo property holds in a number of cases of interest, and has strong implications for other properties of interest.

In practice this problem leads to an attempt to classify the "minimal antichains" in the given class of structures. In the context of permutation patterns a rich set of such antichains has been identified; richer than in the case of tournaments.

This suggests the following problem.

PROBLEM 13. *Give an encoding of permutations as tournaments that allows the minimal antichains of permutation patterns to be interpreted as minimal antichains of tournaments.*

This seems a little too hard as stated, and probably one should settle for an encoding which works for the known minimal antichains. (Preserving minimality seems challenging.)

An example of an undecidable problem with some connections to homogeneity (specifically, the existence of universal graphs with specified constraints) is given in Cherlin [2011b]. One might take this as a hint of a negative solution to Lachlan's problem.

The problem of decidability of j.e.p. for permutation pattern classes determined by finitely many constraints is also open. On the other hand the

---

[1]A recent preprint connects this problem to the study of context-free languages. This is a striking development which presumably is worth looking at model theoretically. See Bodirsky, Knäuer, and Rydval [2021].

homogeneous permutations are known; there are finitely many of them and Lachlan's problem trivializes in that context.

18.2. Standard binary homogeneous structures

The supply of known binary homogeneous structures, while extensive, is not very varied. At some point, after becoming convinced that the conjectured classification of the metrically homogeneous graphs put forward in Part II is reasonable, I began to wonder whether something very similar occurs in general. Phrasing this thought explicitly produces a number of concrete classification problems and a systematic way to look for "natural" examples very broadly.

We define *standard* binary homogeneous structures as follows, and then ask whether the classification of binary homogeneous structures reduces, in a very weak sense, to the standard case.

DEFINITION 18.2.1. Let $\mathcal{A}$ be an amalgamation class of finite binary structures.

1. An *amalgamation strategy* γ for $\mathcal{A}$ is a function on 2-point amalgamation problems P over $\mathcal{A}$ such that $\gamma(P)$ supplies a 2-type which can be used to complete the diagram.

2. A *Henson constraint* relative to γ is a finite binary structure whose 2-types lie outside the range of γ.

3. A *standard* binary homogeneous structure Γ is one whose associated amalgamation class $\mathcal{A}$ has the form

$$\mathcal{A} = \mathcal{A}_3 \cap \mathcal{A}_{\gamma,H}$$

where $\mathcal{A}_3$ is a 3-constrained amalgamation class, γ is an amalgamation strategy for $\mathcal{A}_3$, and $\mathcal{A}_{\gamma,H}$ is the class defined by a set of γ-Henson constraints.

The conjectured classification of metrically homogeneous graphs of generic type from Part II of the previous volume amounts, abstractly, to the conjecture that these structures are standard. Much of Part II is then devoted to figuring out what this means concretely. In this case, γ-Henson usually means $(1,\delta)$-Henson.

PROBLEM 14. *Is there a binary homogeneous structure Γ which satisfies neither of the following conditions?*

(a) *Γ is standard.*

(b) *There is a $a \in \Gamma$ and a non-trivial 2-type p for which*

$$a^p = \{x \mid \mathrm{tp}(a, x) = p\}$$

realizes fewer triangle types than Γ.

As all imprimitive structures fall under case (b), this is a problem about primitive binary homogeneous structures. One would much prefer to have all non-trivial 1-types over a realize fewer triangles, and perhaps in the primitive case this is reasonable. Clause (b) allows for cases like the generic local order which are primitive but not standard.

There is a long-standing conjecture of a considerably more explicit form in the finite case (as we describe below, a proof has now been announced).

CONJECTURE 1. *Every finite primitive binary structure is of one of the following forms.*

(a) Sym_n *acting naturally*;
(b) *A cyclic group of prime order acting regularly*;
(c) *An anisotropic affine orthogonal group over a finite field acting naturally (necessarily of dimension at most 2).*

Note that even finite cliques are standard in our sense (barely): there is an amalgamation strategy which uses no non-trivial 2-type, so any constraint may be considered a Henson constraint. This stretches the notion of free amalgamation and one may prefer to add finite cliques as an explicit exceptional case.

One may check that the affine anisotropic orthogonal groups in dimension 2 also satisfy the stated conditions. In odd characteristic after fixing a point the locus of a 1-type becomes imprimitive with respect to the relation $y = \pm x$. In even characteristic with the order q of the field at least 4 one can check that some 2-type is omitted in one of the induced 1-types. For $q = 2$ we have a finite clique.

Wiscons reduced the proof of Conjecture 1 to the case of primitive actions of almost simple groups in Wiscons [2016], and a systematic attack on the almost simple case by Dalla Volta, Gill, Hunt, Liebeck, and Spiga completed the treatment of all such cases, about the time this book was submitted for publication. (See Appendix B.)

The companion to Problem 14 is the following.

PROBLEM 15. *Classify the 3-constrained binary homogeneous structures explicitly.*

The problem is really to classify the standard ones, but dealing with the 3-constrained case would be the heart of the matter.

In practice, Problem 15 has been part of most censuses which have been undertaken, but has tended to involve isolated examples or the imprimitive case. At present this problem is emerging as a center of attention, accompanied by the question to what extent the solution fits into the framework of generalized metric spaces when the 2-types are all symmetric.

Once one has a grip on the 3-constrained classes, and the associated standard type homogeneous structures, one wants to know, with specific exceptions, that the pattern of forbidden triangles in any homogeneous structure of the same type defines one of the specified 3-constrained classes. In particular, in the case of metrically homogeneous graphs, we know the 3-constrained classes and a full proof of the classification conjecture one would naturally start with this problem: to show that the forbidden triangles in a metrically homogeneous graph of generic type form one of the specified admissible sets of constraints.

Problem 14 includes an inductive element under clause (b); that is, it leads to a classification problem where the structure induced on one of the 1-types over a parameter a is known, and is exceptional.

A related problem is the following.

PROBLEM 16. *Let Γ be a primitive binary structure, $a \in \Gamma$, and p a 2-type. Does it follow that a^p is primitive? Does it follow that the operation of algebraic closure is trivial?*

18.3. Concrete classification problems: overview

We now consider the classification problem for homogeneous 2-multi-tournaments and for homogeneous 3-multi-graphs. Our point of view is that of the preceding section: in other words, we first ask what the 3-constrained homogeneous structures are, and identify their variations (adding Henson constraints). We then ask whether the pattern of forbidden triangles is necessarily as in one of the 3-constrained cases. After that, one wants to work out in each case whether the structures with a specified collection of forbidden triangles are the standard structures associated to that particular collection.

In the case of 2-multi-tournaments and 3-multi-graphs, the 3-constrained homogeneous structures were already identified in the appendix to Cherlin [1998].

Our task in this section is to put the material of Parts I and II into this context, to see what has been accomplished and also what needs to be added at this point to complete the picture. We will devote the rest of the volume to filling in some of the missing pieces.

18.3.1. Classification problems for small languages. The conventional view of the distinction between graphs and tournaments, for a model theorist, is that they share the same language but different axioms. We prefer to view the language as specifying the structure of the set of quantifier-free k-types for some definite value of k (with $k = 2$ here). This structure includes the action of the symmetric group on the variables, as well as the restriction maps to ℓ-types for $\ell < k$. In the case of homogeneous tournaments and graphs, this brings us back to the point of view that tournaments are structures with one pair of anti-symmetric 2-types, and graphs are structures with two symmetric 2-types. We consider only irreflexive k-types (all elements distinct).

The number of irreflexive k-types, counted *up to symmetry*, will be called the *rank* of the language.

From this perspective, here are the known classification results that cover *all* homogeneous structures of a specified combinatorial type in a purely binary language (just one 1-type, and a finite set of irreflexive 2-types, symmetric or anti-symmetric).

- Rank 1, symmetric: Theory of equality.
- Rank 1, anti-symmetric: Tournaments: finitely many.
- Rank 2, symmetric: Graphs: countably many.
- Rank 2, with one symmetric type and one anti-symmetric pair of types: Directed graphs; uncountably many, with one uncountable family, and countably many others.

The complexity of the classification appears to rise quickly and it is unclear whether such explicit classifications can continue much further; but the only definitely known obstacle to this, so far, is the sheer length of the arguments required. It would be remarkable if one could handle all finite binary relational languages using the current methods and patience, but this is merely a more concrete way of phrasing Lachlan's question of effectivity.

We feel that the rank is more significant than the total number of 2-types; that is, while increasing the number of 2-types by breaking symmetry complicates matters substantially, increasing the rank complicates matters even more.

The "next" cases to be considered from this point of view would be the following.

(a) Rank 2, anti-symmetric.

(b) Rank 3, symmetric.

In the terminology we have adopted, structures of the first type are called *2-multi-tournaments*, and structures of the second type are *3-multi-graphs*, in other words tournaments with a coloring of the arcs by two colors, and complete graphs with a coloring of the edges by three colors.

The 3-constrained homogeneous structures of these types were given in Cherlin [1998], with the imprimitive ones and the free amalgamation classes omitted. This was based mainly on a computer search and no documentation was provided there. We will give these lists here, and we supply further details in the following chapters.

18.3.2. Primitive 3-constrained homogeneous 3-multi-graphs without free amalgamation. There is only one primitive 3-constrained homogeneous 3-multi-graph which is not associated with a free amalgamation class. After labeling the 2-types appropriately it may be interpreted as being either of the following metrically homogeneous graphs.

$$\Gamma^3_{3,3,10,11} \text{ or } \Gamma^3_{1,3,8,9}.$$

The forbidden triangles are then of the following types in the two cases.

$$\Gamma^3_{3,3,10,11}: (1, 1, 3), (2, 2, 1), (1, 1, 1);$$

$$\Gamma^3_{1,3,8,9}: (3, 3, 2), (1, 1, 3), (3, 3, 3).$$

The correspondence between these two points of view is given by cyclic permutation of the labels on the 2-types: $(1, 3, 2)$. Notice that the triangle inequality on one side corresponds to a less trivial constraint on the other side.

There are other primitive metrically homogeneous graphs of diameter 3, but the corresponding amalgamation classes have free amalgamation, using distance 2 as the "default" value. (Typically free amalgamation is interpreted as the absence of additional relations, but in binary languages with all 2-types treated on an equal footing, the meaning is that a particular type is to be used to make the amalgam; and it is preferable that the type used be symmetric.)

The infinite imprimitive homogeneous 3-multi-graphs were classified in Cherlin [1999], and all finite homogeneous 3-multi-graphs were classified by Lachlan [1986, §2], with more details given in an unpublished work (Lachlan [ca. 1982]).

We will give the catalog of all known homogeneous 3-multi-graphs in §19.1.

18.3.3. Primitive 3-constrained homogeneous 2-multi-tournaments without free amalgamation. There is a distinctly richer supply of primitive 3-constrained 2-multi-tournaments which do not correspond to free amalgamation classes. This is shown in Table 18.1, p. 10, which shows the (non-degenerate) possibilities up to a permutation of the language. The table has been arranged and labeled to suit our present purpose. We use the labels 1, 2 for the two colors of arc, and we list the forbidden triangles using the following conventions.

NOTATION 18.3.1. For $i, j, k \in \{1, 2\}$, the symbol "$C_3(i, j, k)$" denotes an oriented 3-cycle with arcs of type (i, j, k) respectively: $a \xrightarrow{i} b \xrightarrow{j} c \xrightarrow{k} a$ (so this is also denoted $C_3(j, k, i)$ and $C_3(k, i, j)$).

Similarly, $L_3(i, j, k)$ is a transitive tournament on three vertices a, b, c where $a \xrightarrow{i} b \xrightarrow{j} c$ and $a \xrightarrow{k} c$.

When listing forbidden triangles we use a compressed notation, e.g., "C_3 : 111, 112" stands for "$C_3(1, 1, 1)$, $C_3(1, 1, 2)$."

Note in particular that when there is a definable linear order it may be taken to be given by $\xrightarrow{1} \cup \xrightarrow{2}$; there are four such cases, numbered 2–5. These are the cases in which the structure may be viewed as an ordered graph, or as an ordered tournament.

This catalog will be discussed in Chapter 22. The label "Exceptional" means "poorly understood" and, in particular, the existence of those structures is one of the points we will need to check.

#	Constraints		Type
	Group I: Finite		
1	C_3: 111,112,222	L_3: 111,121,122,211, 212,221,222	Pentagram
	Group II: Linearly ordered by $\xrightarrow{1} \cup \xrightarrow{2}$		
	Common Constraints		
	C_3: 111,112,221,222	L_3: None	
	Additional Constraints		
2	C_3: none	L_3: none	RG $*<$
3	C_3: none	L_3: 112	$\leq$ extends PO
4	C_3: none	L_3: 111	Henson $*<$
5	C_3: none	L_3: 112,221	$<*<$
	Group III: p.o. by $\xrightarrow{1}$, no definable linear order		
	Common Constraints		
	C_3: 111, 112	L_3: 112	
6	C_3: none	L_3: none	Desymm. PO
7	C_3: 221	L_3: none	Exceptional
	Group IV: Infinite, no definable partial order		
	Constraints		
8	C_3: 111,112	L_3: none	Exceptional
9	C_3: 112	L_3: 111	"
10	C_3: 111,112	L_3: 111	"
11	C_3: 111,222	L_3: 111,122,212	$\widetilde{\mathbb{S}}(3)$
12	C_3: 111,112	L_3: 121,211,221,222	$\mathbb{S}(4)$

TABLE 18.1. Primitive 3-constrained non-degenerate homogeneous 2-multi-tournaments without free amalgamation.

18.3.4. The story so far: 3-constraint. Recall the standard plan of attack for classification theorems in binary languages.

1. Determine the 3-constrained structures.
2. Determine the associated standard structures.
3. Attempt to show that the set of forbidden triangles in a homogeneous structure taken by itself already determines an amalgamation class, on the list above. Note any exceptions (aiming for a complete classification of all such at this stage).
4. Attempt to show, ideally, that every homogeneous structure in the class is standard, in the sense corresponding to the associated 3-constrained

structure. When this fails, aim to show that one of the structures induced on the locus of some non-trivial 1-type over a point was previously classified, and complete the classification on a more ad hoc basis.

For example, in the classification of homogeneous tournaments, the treatment of the generic local order falls under point (4), when no triangle is forbidden but the structure is not generic.

The lists given in Cherlin [1998, Appendix] omit the imprimitive and free amalgamation cases, but we will want to include them here. The free amalgamation classes present no difficulties from this point of view but the imprimitive case requires its own separate treatment.

As mentioned above, the imprimitive homogeneous 3-multi-graphs are known. On the other hand, the imprimitive homogeneous 2-multi-tournaments were never classified. We will fill this gap in §20.3, as follows.

PROPOSITION 20.3.2. *Let Γ be an imprimitive homogeneous 2-multi-tournament. Then up to a change of language Γ is one of the following.*

- *A composition $T_2[T_1]$ with T_i an $\xrightarrow{i}$-tournament.*
- *(Shuffled type) Γ is derived from a homogeneous local order $\Gamma_1 \cong \mathbb{Q}$ or $\mathbb{S}$ which is partitioned into n dense pieces, with $2 \leq n \leq \infty$, and with a $\xrightarrow{2} b$ in Γ iff a, b lie in distinct pieces of Γ_1, and a $\xrightarrow{1} b$ holds in Γ_1.*
- *(Semi-generic type) Generic imprimitive with infinite components (type $\mathbb{Q}$, $\mathbb{S}$, or Γ^∞) and satisfying the* parity constraint *described below.*
- *(Generic type) The generic de-symmetrization of an imprimitive homogeneous directed graph $n \cdot T$ with $2 \leq n \leq \infty$, where T is an infinite homogeneous tournament.*

We will leave the precise definitions used above to §20.3. The proof of the proposition follows Cherlin [1987], at a greater level of generality. The original argument strikes us as very compressed at points and possibly incomplete.

Thus both the imprimitive homogeneous 3-multi-graphs and the imprimitive homogeneous 2-multi-tournaments are classified explicitly, and we may set these aside in any further analysis of these structures.

There is no difficulty in verifying the accuracy of the list of 3-constrained 3-multi-graphs and in fact we will arrive at a more general result in Proposition 19.2.1, quoted below.

On the other hand, the classification of 3-constrained homogeneous 2-multi-tournaments is both richer and harder to verify, both in terms of the existence of the structures (the amalgamation property) and the completeness of the list. We deal with this in §21.1.

PROPOSITION 21.1.1. *Up to a permutation of the language, the primitive 3-constrained homogeneous 2-multi-tournaments are the twelve shown in Table 18.1 together with the free amalgamation classes, which forbid only $C_3(1,1,1)$, $L_3(1,1,1)$, or both.*

18.3.5. Forbidden triangles in homogeneous 3-multi-graphs, and beyond. The next question in logical order is whether the set of forbidden triangles for a given homogeneous structure in our class is one of the known patterns defining a 3-constrained homogeneous structure, and if not, what exceptions arise. Only after this would we come to the question as to whether, in each case, the structures associated with one such pattern of forbidden triangles are in fact standard, a point settled in certain cases by Amato, Cherlin, and Macpherson [2021] or Part I of this monograph.

In the case of homogeneous 3-multi-graphs, the cases treated in Amato, Cherlin, and Macpherson [2021] can be used to complete the analysis of forbidden triangles for homogeneous 3-multi-graphs in general. Thus in Chapter 19 we show that the following proposition is an easy consequence of the known classification results.

PROPOSITION 19.2.1. *Let Γ be a homogeneous 3-multi-graph not in the catalog of known examples. Then any triangle omitted by Γ is monochromatic, and the set of triangles omitted by Γ defines a free amalgamation class.*

From this point onward, the task is to show that any homogeneous 3-multi-graph not covered by the prior results corresponds to a free amalgamation class.

It is not clear whether one should expect that this will actually be the case; and even if it is the case, it is very unclear how to adapt previous arguments to prove it. This is the most significant question arising within this classification problem, as there is a realistic possibility that examples of a new type will arise. But some cases of this type were handled in Amato, Cherlin, and Macpherson [2021] and did in fact lead to free amalgamation classes.

We continue the analysis somewhat further, in a direction suggested by clause (b) of Problem 14. If a monochromatic triangle of type i is forbidden, then the corresponding 1-type over a parameter does not realize the type i and may therefore be viewed as a homogeneous graph. One particular case that should be eliminated is that in which the induced structure is imprimitive. We refer to this situation as the *locally degenerate imprimitive case* since the induced structure on a 1-type is both degenerate (omits a 2-type) and imprimitive.

We will show the following.

PROPOSITION 19.3.44. *Suppose that Γ is a homogeneous 3-multi-graph which is locally degenerate imprimitive. Then Γ is of known type.*

In general, the case in which there is some forbidden clique (monochromatic subgraph) is a candidate for inductive analysis, with the case above arising at the base of the induction. Of course, that is a very broad case and would be expected to require the style of analysis of Part I, at a minimum.

18.3.6. Forbidden triangles in homogeneous 2-multi-tournaments. One would expect a language with two pairs of asymmetric 2-types to be easier to handle than one with three symmetric 2-types, but this is certainly not the case as far as the analysis of patterns of forbidden triangles is concerned. The relevant combinations of forbidden triangles, even in the 3-constrained case, are more varied, but even the analysis of individual cases turns out to be more complex.

So a central concern in Chapters 21 and 22 will be the allowable patterns of triangles in a general homogeneous 2-multi-tournament.

18.3.6.1. *The 3-constrained case.* To begin with one classifies the 3-constrained cases; this is the topic of Chapter 21. The result is shown in Table 21.1, §21.1. There we restrict ourselves to cases with no $\emptyset$-definable linear order, as Part I of this monograph gives the full classification in the presence of a linear order. This provides a substantial simplification: of the twelve 3-constrained classes not associated with free amalgamation, one is finite and four involve $\emptyset$-definable linear orders, so only seven are left, along with the free amalgamation classes shown for the sake of completeness but playing no real role in our present concerns.

Two of these seven cases allow a $\emptyset$-definable partial order, which in this particular context means only that one of the two 2-types is transitive. Of the remaining five, two have natural interpretations as being either generalizations of the generic local order or closely related to such a generalization, while the other three have not been given a natural interpretation as yet. However two of them arise as the 3-constrained structures within a particular infinite family of homogeneous 2-multi-tournaments.

18.3.6.2. *The general case.* We turn to the problem of the identification of the possible *patterns of forbidden triangles* in infinite primitive homogeneous 2-multi-tournaments with no $\emptyset$-definable linear order, where this pattern is not associated with any free amalgamation class. Then our target is one of the patterns associated with the 3-constrained classes found, and specifically those associated with entries numbered 6–12 in the list given in Table 21.1.

One learns from the classification of the 3-constrained classes that in four cases the relevant amalgamation diagrams have order 6. We focus here on the points which can be checked in the 3-constrained case using diagrams of order at most 5. This leads to the following result.

PROPOSITION 22.1.1. *Let Γ be an infinite, primitive, homogeneous 2-multi-tournament not associated with a free amalgamation class. If Γ has a $\emptyset$-definable linear order then it is found in the classification in Part I. If not, then either the set of forbidden triangles in Γ defines one of the known 3-constrained 2-multi-tournaments, or one of the following four cases applies.*

1. *Triangle types $C_3(1, 1, 1)$ and $C_3(2, 2, 2)$ are forbidden and all other triangle types are realized.*

2. *Triangle types $C_3(1, 1, 1)$ and $L_3(2, 2, 1)$ are forbidden and all other triangle types are realized.*

3. *Triangle type $C_3(1, 1, 2)$ is forbidden and all other triangle types realized.*

4. *Triangle types $C_3(1, 1, 2)$ and $L_3(2, 2, 1)$ are forbidden and all other triangle types realized.*

It seems likely that these four remaining cases, corresponding to amalgamation diagrams of order 6 in the 3-constrained case, will be challenging to treat in general, but they should eventually be eliminated by the same sort of direct analysis.

This analysis is given in Chapter 22 and is summarized in §22.8.

18.3.7. The catalog revisited. We conclude in §22.9 with an overview of the results achieved on forbidden triangles in homogeneous 2-multi-tournaments, as well as the results known only in the 3-constrained case. Our final Tables 22.1 (imprimitive case) and 22.2 (primitive case) contained a detailed and explicit catalog of the known homogeneous 2-multi-tournaments, with the constraints of order 3 listed, and other supplementary constraints indicated where required.

Chapter 19

HOMOGENEOUS 3-MULTI-GRAPHS

The results to be proved in this chapter, and succeeding chapters, have been presented in detail in Chapter 18.

Namely, our subject in the present chapter is the classification problem for homogeneous 3-multi-graphs, in the terminology adopted in Chapter 18. These are complete edge labeled graphs with label set $\{1, 2, 3\}$; in this language, 2-*multi-graphs* are essentially ordinary graphs, after deciding that one of the two colors represents edges and the other non-edges. Similarly homogeneous graphs can be interpreted as homogeneous 3-multi-graphs in six different ways, using two of the three colors. These are called *degenerate* 3-multi-graphs, as one 2-type is omitted.

As mentioned in Chapter 18, the finite or imprimitive homogeneous 3-multi-graphs have been classified previously.

This chapter will also rely on the main result of Amato, Cherlin, and Macpherson [2021], showing that the conjectured classification of metrically homogeneous graphs is valid in diameter 3. The proof of that result itself makes occasional use of Part II of the present monograph. So we can view that work as a bridge from Part II of the previous volume to the present volume.

We first give a full catalog of the known examples of homogeneous 3-multi-graphs. Then we determine the patterns of forbidden triangles which may occur in an arbitrary homogeneous 3-multi-graph not covered by one of the prior classification results (Proposition 19.2.1). At that point we see that the result of Amato, Cherlin, and Macpherson [2021] brings us down to the generic case of the classification problem for homogeneous 3-multi-graphs, by which we mean the case in which the target consists exclusively of structures associated with free amalgamation classes (classes closed under one of the relevant notions of free amalgamation).[2] This reduction goes quickly, modulo the previously known results.

This generic case divides further into a more special case in which some clique is forbidden, where induction on the size of the forbidden clique is helpful,

---

[2] We used the term *generic type* in a different sense in Part II, specific to the context of metrically homogeneous graphs.

and the case in which no clique is forbidden, which provides an interesting challenge; a similar challenge is dealt with in Part I of this monograph, but in a considerably simpler context.

As we have seen in this monograph, it is useful to examine the local structure of homogeneous structures; in the case of 3-multi-graphs, these are the structures $\Gamma_1, \Gamma_2, \Gamma_3$ induced on the realizations of one of the 1-types over a basepoint (labeled by the corresponding 2-type over the empty set). In particular one has the exceptional case in which Γ_1 (say) is imprimitive. We will consider the case in which Γ_1 is not only imprimitive, but degenerate. We refer to this as the *locally degenerate imprimitive* case. This takes quite a long analysis to dispose of (§ 19.3, in seven parts). One recognizes a certain style of argument associated with local analysis but for the present the method involves a close examination of various concrete possibilities. Since these are all eliminated in due course, it is reasonable to seek more general methods, applicable when the full structure is primitive.

It would have been more satisfactory to dispose of the locally imprimitive case in full, but issues of both time and space finally weighed in. We return to a summary of the state of affairs with respect to the classification of homogeneous 3-multi-graphs at the end of the chapter (§19.4).

19.1. A catalog

We give a catalog of the known homogeneous 3-multi-graphs. We fix notation for the irreflexive 2-types: these will be denoted by $\frac{1}{\quad}$, $\frac{2}{\quad}$, and $\frac{3}{\quad}$, or more simply by $1, 2, 3$ where possible.

The *degenerate case* in which not all 2-types occur is covered by the Lachlan/Woodrow classification. We do not include this below.

We first give a classification of the examples into eight families, six of them imprimitive. We will give an explicit structural description of each family below.

1. Imprimitive
   (a) Composite
   (b) Product
   (c) Double cover
   (d) Clique-restricted
   (e) Semi-generic
   (f) Generic imprimitive
2. Primitive
   (a) $\Gamma^3_{K_1, K_2, C_0, C_1, \mathcal{S}}$ with admissible parameters
   (b) Free amalgamation class

We now describe these families in detail. They are intended to be non-overlapping, and generally speaking we may take it as tacitly given that each family includes the condition of not falling into an earlier family. But some marginal cases fall naturally under more than one heading.

1(a) Composite. The characteristic feature is that *there is a non-trivial congruence* (i.e. the type of a pair of points is determined by their equivalence classes).

The 3-multi-graph has the form $G_2[G_1]$ with G_1, G_2 homogeneous graphs in disjoint languages. The automorphism group is a wreath product.

As our language allows three 2-types, one of the factors G_1, G_2 is complete. The other factor may itself be composite, in which case the structure has the form $K[K[K]]$ with each K an $\overset{i}{\text{—}}$-clique, for varying $i \in \{1, 2, 3\}$.

In the remaining imprimitive cases, the equivalence classes for any minimal equivalence relation must be cliques of some type (i.e., monochromatic).

1(b) Product. The characteristic feature is *a pair of transversal equivalence relations*.

Like composition, the product construction $G_1 \otimes G_2$ can be defined in great generality.

The vertex set is the underlying Cartesian product, and the type of the pair $(a_1 b_1, a_2 b_2)$ is the pair of types $(\text{tp}(a_1 a_2), \text{tp}(b_1 b_2))$. The automorphism group is the direct product of the automorphism groups of the factors.

In general, if G_1, G_2 have k and ℓ non-trivial 2-types, then $G_1 \otimes G_2$ has $k\ell + k + \ell$ non-trivial 2-types. If the product has three non-trivial 2-types, this means that both G_1 and G_2 must be complete. So this is the case of interest here. The product of two complete graphs is a grid with three non-trivial 2-types: equal first coordinate, equal second coordinate, or neither. Or in more geometric terms: vertically related, horizontally related, or in general position.

On the other hand, the composition of two complete graphs $K_m[K_n]$ is a complete multipartite graph with m parts of size n; or its complement, the disjoint union of m copies of K_n, depending on which 2-type is called an edge, and which is called a non-edge.

1(c) Double cover. The characteristic feature is *an equivalence relation with classes of order 2, which is not a congruence, and preferably not a product.*

Construction: Let G be a homogeneous graph, and identify the two non-trivial 2-types (edge, non-edge) with the labels $1, 2$ respectively.

Form an extension $G^* = G \cup \{\star\}$ with $\star$ adjacent to every vertex of G. Then take two copies G_1^*, G_2^* of G^*, and set $\hat{G} = G_1^* \cup G_2^*$ with edges and non-edges (i.e., types 1 and 2 respectively) between G_1^* and G_2^* corresponding to non-edges and edges (types 2 and 1 respectively) within G^*, apart from the pairs (a, a') corresponding to the same element of G^*, which realize the third type $\overset{3}{\text{—}}$.

Note that this construction results in the following properties. If the point $*_1$ in G_1^* is taken as basepoint, we get $\Gamma_1 = G_1$, $\Gamma_2 = G_2$, $\Gamma_3 = \{*_2\}$, and we also have a kind of "antipodal" law $\mathrm{tp}(a_1, b_2) = 3 - \mathrm{tp}(a_1, b_1)$ if we consider equality as type 0. Examples follow below.

The result of this construction is not always homogeneous. But if the automorphism group of the double cover is transitive, then the structure is homogeneous. These are in fact metrically homogeneous graphs of diameter 3 and antipodal type (apart from the disconnected case (a) below). There are four types of double cover, if we include under this heading a case which was already treated as a product.

(a) With G a clique, and thus G^* also a clique, the double cover is $K_2 \otimes G^*$.

(b) With G the pentagon, the double cover is the 1-skeleton of the icosahedron.

(c) With G the primitive non-degenerate homogeneous graph of order 9, the double cover is a finite antipodal graph of order 20. Cameron's presentation of this graph is as the graph of 3-subsets of a 6 element set, with edge relation given by

$$|x \cap y| = 2.$$

(d) With G the random graph, the double cover is the generic antipodal metrically homogeneous graph of diameter 3.

1(d) Clique-restricted. We fix a 2-type $\overset{i}{-}$ relating some pair of inequivalent points. In this case, we add the constraint that an i-clique of some fixed order m is forbidden. We also fix the number n of equivalence classes (which may be infinite) subject to $n \geq m$.

There is a generic graph of this type for each allowable choice of m, n.

1(e) Semi-generic. Here one requires that for any two pairs of points taken from distinct equivalence classes, the number of edges of a given type between the two pairs be even.

There is a generic graph of this type (that is, the associated class of finite structures is an amalgamation class, hence has a Fraïssé limit).

1(f) Generic imprimitive. Finally, we have a graph with a specified number n of equivalence classes, with $2 \leq n \leq \infty$, and otherwise generic. The induced structure on each class is a clique, and any two classes constitute a generic bipartite graph. This does not exhaust the properties of the graph, but fixes the ideas.

This completes the discussion of imprimitive families. That part of the catalog is complete: an imprimitive homogeneous 3-multi-graph lies in one of these families.

We now review the two known primitive types.

2(a) Primitive metric; 3-constraints and Henson constraints. We have met the graphs $\Gamma^3_{K_1, K_2, C_0, C_1, \mathcal{S}}$ in Part II.

2(b) Free amalgamation classes. In the case of free amalgamation, we take any set $\mathcal{C}$ of finite 3-multi-graphs with the property that at most two 2-types are involved in all of the constraints. Taking $\mathcal{C}$ as forbidden, we have free amalgamation with respect to the third 2-type, and a corresponding generic 3-multi-graph $\Gamma_{\mathcal{C}}$.

As we have mentioned several times, this catalog of known types has been proved complete in the following three cases.

(a) In the finite case, it is treated by a direct inductive analysis in Lachlan [1986, §2], with more details given in Lachlan [ca. 1982].

(b) In the imprimitive case, it is covered by Cherlin [1999].

(c) In the metrically homogeneous case, it is treated in Amato, Cherlin, and Macpherson [2021].

Note that the metric case, in the setting of 3-multi-graphs, is simply the case in which triangles of type $(1, 1, 3)$ are forbidden. This does not appear to be a very striking condition—as such, it defines a free amalgamation class—but it turns out to simplify the problem enormously.

We sum all this up as follows.

Remark 19.1.1. Any homogeneous 3-multi-graph not in the above catalog (up to a change of language) is infinite and primitive.

We will put this remark in a sharper form in the next section.

19.2. Triangle constraints

In this section we aim at the following.

PROPOSITION 19.2.1. *Let Γ be a homogeneous 3-multi-graph not in the catalog given above. Then any triangle omitted by Γ is monochromatic, and the set of triangles omitted by Γ defines a free amalgamation class.*

As we will see, the bulk of the proof is covered by the main result of Amato, Cherlin, and Macpherson [2021].

NOTATION 19.2.2. Let Γ be a homogeneous 3-multi-graph.
For any vertex $v \in \Gamma$ and any non-trivial 2-type p we set

$$v^p = \{x \in \Gamma \mid \mathrm{tp}(vx) = p\}.$$

When p is the type $\overset{i}{-}$ we also write v^i for v^p. Explicitly:

$$v^i = \{x \in \Gamma \mid v \overset{i}{-} x\}.$$

In particular, we fix an arbitrary basepoint denoted by v_* and we define the homogeneous 3-multi-graphs Γ_i correspondingly:

$$\Gamma_i = v_*^i.$$

These are the *local* constituents of Γ, and are again homogeneous 3-multi-graphs.

We denote by $\overset{i}{\sim}$ the binary relation which is the reflexive extension of $\overset{i}{—}$, which is of particular interest when it is an equivalence relation on Γ, or, indeed, whenever its restriction to one of the local constituents Γ_j is an equivalence relation.

More generally, if S is a set of 2-types, $\overset{S}{\sim}$ denotes the reflexive extension of their union.

In our diagrams we represent type 1 by a solid edge, type 2 by a dotted edge, and type 3 by a stippled edge with wide stipples.

LEMMA 19.2.3. *Let* Γ *be a homogeneous 3-multi-graph in which triangles of type* (i, i, j) *are forbidden for some fixed and distinct* $i, j \in \{1, 2, 3\}$. *Then* Γ *is in the catalog above.*

PROOF. We may assume that triangles of type $(1, 1, 3)$ are forbidden. We will consider the graph structure on Γ with edge relation $\overset{1}{—}$ (but Γ retains its original language).

We know that Γ is primitive. So the graph on Γ is connected. By Fact 1.4.1 (an elementary result, from Chapter 1 of Volume I) the metric on Γ is the path metric, and hence Γ is a metrically homogeneous graph. Now the classification in Amato, Cherlin, and Macpherson [2021] applies.   □

LEMMA 19.2.4. *Let* Γ *be a homogeneous 3-multi-graph in which triangles of type* $(1, 2, 3)$ *are forbidden. Then* Γ *is in the catalog above.*

PROOF. By the previous lemma we may assume that all triangles of type (i, i, j) with i, j distinct are realized in Γ. We now sharpen this as follows.

CLAIM 1. *For any* i, j *distinct, triangles of type* (i, j, j) *are realized in* Γ_i.

We may take $i = 1$, $j = 2$, so that we aim to put a triangle of type $(1, 2, 2)$ into Γ_1.

We first consider the following amalgamation diagram, in which the three paths from a_1 to a_2 are colored using all three possible pairs of colors.

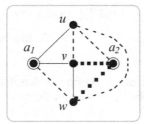

Since any completion would contain a triangle of type $(1, 2, 3)$, this is a diagram with no completion, and so one of the factors does not embed in Γ.

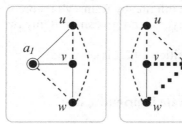

The factor omitting a_1, shown on the right, is forced by the amalgamation in which (u, v) are treated as a pair whose type is to be determined. The factors of this latter diagram are triangles afforded by the previous lemma.

So the factor omitting a_2, shown on the left, is omitted. Therefore, if we treat this factor as a diagram in which the type of (u, v) is to be determined, the unique completion has $u \overset{1}{\relbar} v$, as shown.

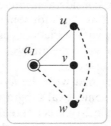

With v as basepoint, this gives a triangle of type $(1, 2, 2)$ in v^1, proving our claim.

Now consider the following amalgamation diagram.

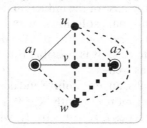

The factors of this diagram consist of a triangle of type $(1, 1, 2)$ in w^2 and a triangle of type $(2, 3, 3)$ in u^2, hence embed in Γ by the claim. But the diagram has no completion. This contradiction proves the lemma.

$\square$

Proof of Proposition 19.2.1. By Propositions 19.2.4 and 19.2.3 a new homogeneous 3-multi-graph Γ must realize all triangle types other than the monochromatic triangles, i.e., those of type $(1, 1, 1)$, $(2, 2, 2)$, or $(3, 3, 3)$. Furthermore, as Γ must be infinite, by Ramsey's theorem we must realize at least one of these triangle types, and we can omit any two of them in a free amalgamation class.

Thus the proof of Proposition 19.2.1 is complete.   □

19.3. The locally degenerate imprimitive case

Local analysis is the study of the local constituents of a homogeneous structure. As we saw in Part II this can be developed extensively in the case of metrically homogeneous graphs and can then be usefully applied to obtain concrete classification results. Generally speaking, a classification conjecture predicts what constituents will occur, and the extent to which their structure reflects the structure of the whole. In particular in the primitive case we expect the local constituents to be primitive.

Accordingly, a special case of considerable interest at the outset is the *locally imprimitive case*, that is, the case in which for some ℓ the local constituent Γ_ℓ is imprimitive. This covers a very wide variety of cases which we would hope to eliminate at a relatively early stage of the analysis, eventually clearing the way for a robust analysis of the generic case.

In this section we are concerned with the *locally degenerate imprimitive case*, meaning that not only is some local subgraph imprimitive, but it is also degenerate in the sense that some 2-type is not realized. In other words we are assuming here that a triangle of type (i, i, j) is forbidden, and setting aside the known cases, this means that $i = j$ and an i-clique of order 3 is forbidden. More generally, the case in which some i-clique is forbidden lends itself to induction on the order n of the forbidden clique. In some sense the case $n = 2$, the degenerate case, is the true base of the induction, but since it will come with many "sporadic" or "exceptional" solutions, in an inductive treatment these will all occur, potentially, as local constituents. We do not take up the locally imprimitive non-degenerate case; it is possible that some of our arguments would go over to that case as well.

In the case of 3-multi-graphs, an imprimitive and degenerate local constituent must have the form $K[K]$, or more explicitly $K_m^i[K_n^j]$, that is a composition of cliques of orders m, n and types i, j, respectively, where i, j are distinct and $m, n \geq 2$; here we allow m or n to be infinite.

We aim here at the following.

Proposition 19.3.44. *Suppose that Γ is a homogeneous 3-multi-graph which is locally degenerate imprimitive. Then Γ is of known type.*

Remark 19.3.1. Up to a permutation of the language, the known locally degenerate imprimitive homogeneous 3-multi-graphs are the following.

(a) (Composite) $K_2^1[\Delta]$ with Δ an imprimitive homogeneous graph.

(b) The generic 1-triangle free imprimitive 3-multi-graph with equivalence relation $\overset{2}{\sim}$, with $n \geq 3$ classes

In particular these 3-multi-graphs are imprimitive, and since all imprimitive homogeneous 3-multi-graphs are known, another way to phrase Proposition 19.3.44 is as follows: there is no homogeneous 3-multi-graph which is both primitive and locally degenerate imprimitive.

As we have said, Proposition 19.3.44 is a natural first step toward the anticipated classification of homogeneous 3-multi-graphs omitting a monochromatic triangle. We will return to this at the end.

The proof of this result will require a detailed analysis of a number of special cases, so we first set out some notations and assumptions that will remain in force throughout.

Since we are only interested in cases not already in the catalog of known examples, we may suppose that the locally degenerate imprimitive homogeneous 3-multigraph Γ is itself primitive and infinite, and realizes all triangle types which are not monochromatic.

NOTATION 19.3.2. We choose notation so that Γ_1 is degenerate and imprimitive. Since only monochromatic triangles may be omitted, it follows that Γ_1 omits the 2-type $\overset{1}{-}$; that is, Γ omits a triangle of type $(1, 1, 1)$.

Recall that $\overset{i}{\sim}$ denotes the reflexive extension of $\overset{i}{-}$. We may suppose that the non-trivial $\emptyset$-definable equivalence relation on Γ_1 is $\overset{2}{\sim}$. That is, $\Gamma_1 \cong K_m^3[K_n^2]$ for some m, n with $2 \leq m, n \leq \infty$. And by homogeneity, the same applies to u^1 for any vertex $u \in \Gamma$.

Remark 19.3.3. $\max(m, n) = \infty$. In other words, Γ_1 is infinite.

This holds since Γ is infinite, primitive, and $\aleph_0$-categorical. It follows that the algebraic closure of a point, in the model theoretic sense, is trivial, and hence each local constituent is infinite.

19.3.1. Some general ideas. We will need to examine a large number of special configurations and apply homogeneity (or amalgamation) in a few different ways. We digress here to build up some intuition which may be helpful to the reader going through these arguments in detail.[3]

From a purely technical point of view, one method for eliminating the more recalcitrant configurations is to consider all the defining conditions for the configuration as defining a class of finite structures, and to work out why this is

---

[3]This topic is much more richly illustrated by the arguments of Chapter 15, but we do not assume any familiarity with that material.

not already an amalgamation class, which may lead to some insight or at least to a correct argument. As one approaches the conclusion that the structure under consideration is behaving like the Fraïssé limit of a class that does not have amalgamation, this style of argument becomes increasingly likely; see for example the proof of Proposition 19.3.35, or its immediate predecessor (in order of proof) Lemma 19.3.38.

But in practice one can often give a more transparent argument based on a few simple principles, one of them being the fact that we have only three nontrivial 2-types at our disposal, and complex configurations may produce more than that; or alternatively some simple structure which is not homogeneous for a binary language may embed in some useful way into a more complex structure.

The following principle is helpful, and allows of some further variations.

LEMMA 19.3.4. *Let Γ be a homogeneous structure for a finite relational language, A a finite subset, and B_1, B_2 the sets of realizations of two 1-types over A. Suppose that there is an A-definable bijection $f : B_1 \to B_2$. Then up to a change of language f is an isomorphism. In particular B_1 and B_2 realize the same number of types.*

PROOF. What this means, concretely, is that the type of a sequence in B_1 determines, and is determined by, the type of the corresponding sequence in B_2. Indeed, treating A as a set of constants, if b is a sequence of elements in B_1, then the type of $f[b]$ in B_2 can be viewed as part of the type of b in B_1. □

More generally, if the map f in question is only an A-definable map from B_1 to B_2, it must in any case be surjective. If B_1 is primitive, then f will automatically be injective (unless B_2 consists of a single element and f is constant). We may extend further to the case in which f is a bijection between a quotient of B_1 and a quotient of B_2; if these quotients are taken with respect to congruences then these quotient structures are again homogeneous, and one can work either syntactically or directly with the relevant automorphism groups.

We are interested in certain key failures of homogeneity, since these serve as a template for various proofs by contradiction. We give two closely related examples.

Example 19.3.5. Fix $k \geq 2$ finite and take a set of $n \geq 2k$ points A. View the k-subsets of A as the elements of a structure equipped with binary relations $R_i(a, b)$ defined by

$$|a \cap b| = i$$

for $0 \leq i < k$. For example, for $k = 2$ and $n = 5$ this is the Petersen graph. This structure is not homogeneous: we can find triples of k-sets (a_1, a_2, a_3) or (b_1, b_2, b_3) such that pairwise intersections $a_i \cap a_j$ or $b_i \cap b_j$ all have cardinality $k - 1$, for i, j distinct, but $|a_1 \cup a_2 \cup a_3| = k + 1$ and $|b_1 \cup b_2 \cup b_3| = k + 2$.

The automorphism group of this structure is $\mathrm{Sym}(n)$ acting naturally, so these two triples have distinct types, but they involve a unique 2-type.

In the case $k = 2$ the vertices may be thought of edges in a complete graph, with edge and non-edge relations corresponding to disjoint or non-disjoint edges, and the relevant triples are then the edges of a triangle or the edges of a star, the point being that these are not isomorphic but when reduced to any two edges give a path of length 2.

This example shows up in the proof of Lemma 19.3.25, point (4), and similarly in the proofs of Proposition 19.3.26, Claim 1, of Lemma 19.3.29, Claim 1, and of Lemma 19.3.30, point (1).

An elaboration on this train of thought is found in the proof of Lemma 19.3.8, Claim 1, in a more elaborate context in which points are correlated with certain k-sets (whose elements are points in a quotient structure), but are not necessarily uniquely determined by them.

Example 19.3.6. We consider homogeneous bipartite graphs. It is natural to view the bipartition as given either by an equivalence relation with two classes, or by a pair of unary predicates. We consider the latter possibility, which allows for more examples as the automorphism group need not act transitively.

There are very few homogeneous bipartite graphs, which goes back to the fact that some very simple configurations are incompatible with homogeneity, cf. Goldstern, Grossberg, and Kojman [1996] (this reference deals also with the uncountable case, which we set aside).

Namely, up to bipartite complementation, the homogeneous bipartite graphs are the following.

(a) Complete bipartite;
(b) Perfect matching (in particular, the parts have the same cardinality);
(c) Generic (homogeneous universal) or "random."

The crux of the matter is that if a point on one side has finitely many neighbors, then it either has at most one neighbor, or at most one non-neighbor, after which the analysis is trivial. It will be instructive to see why this is the case. So let the vertices on side A have precisely k neighbors on side B, and let $|B| = n \leq \infty$, where k is finite. Passing to the bipartite complement if necessary we may suppose that $k \leq n/2$. We then have a definable map from the set A to the structure considered in the previous example. By homogeneity it is surjective. As the structure on A is trivial, A is primitive, and therefore this map is either 1-1 or constant. As the map is surjective, it must be 1-1. So we have a definable bijection between A and the k-subsets of B. Now we may conclude in a number of ways: the set A has a unique non-trivial 2-type, and for $k \geq 2$ this does not hold for the k-sets, which quickly gives a contradiction. A more sophisticated argument is this: the k-sets do not form a homogeneous structure for a binary language, and A does, so the definable bijection quickly leads to a contradiction.

The philosophy that this suggests is that when finite parameters come up in the context of homogeneous structures, they tend to be 0 or 1 (at least, up to a change of notation). Of course this is not at all true when the structures themselves are finite, but the principle is reasonable in the context of infinite primitive structures for finite binary languages.

Another canonical example of failure of homogeneity in a very rudimentary structure occurs in the proof of Lemma 19.3.17, which we give in some generality when we come to it.

Example 19.3.7. One sees from the classification of the homogeneous graphs that the graphs $K_n \square K_n$ are not homogeneous for $n \geq 4$; and since the 2-types in these graphs reduce to the edge and non-edge relation, this means that they are not homogeneous in any binary language, which is a more useful point of view for us.

These may be thought of in several ways. They are the line graphs of complete bipartite graphs $K_{n,n}$. In terms of their automorphism groups they correspond to the wreath ptroduct $\mathrm{Sym}(n) \wr \mathrm{Sym}(2)$ with its product action. More concretely, $K_n \square K_n$ is the graph product with vertex set $[n] \times [n]$ and with edge relation $(i, j) - (i', j')$ defined by "$i = i'$ or $j = j'$" for distinct vertices (i, j) and (i', j').

In these graphs, the 3-types correspond to isomorphism types of subgraphs, but the 4-types do not: in terms of the underlying grid structure, edges may be horizontal or vertical, and a pair of edges in general position will give isomorphic subgraphs, but the edges may be parallel or not, corresponding to distinct 4-types.

In addition, we point to the content of Lemma 19.3.12 below as an elementary and useful consequence of homogeneity, one which is a little more technical (or at least specialized), but relevant to the study of locally imprimitive homogeneous 3-multi-graphs generally. We will not elaborate on it here, as this is best illustrated by the analysis that begins with Lemma 19.3.12.

Now we return to our main line of argument.

In the next three sections we eliminate the cases in which m or n is finite. The remaining case, $m = n = \infty$, presents the main difficulty.

19.3.2. The case $m < \infty$: preliminary analysis. We first eliminate the case in which m is finite (and thus n is infinite), in several steps (and in two sections). The goal of the preliminary analysis is Lemma 19.3.19 (page 38), giving considerable information about the 1-types realized in Γ_2 over Γ_1.

The statement of the next lemma allows for one special configuration which is superficially plausible but does not actually occur; but we leave the verification of that point for the lemma immediately following it.

LEMMA 19.3.8. *Suppose that Γ is a locally degenerate imprimitive homogeneous 3-multi-graph with $\Gamma_1 \cong K_m^3[K_\infty^2]$ and $m < \infty$, and that Γ realizes all triangle types which are not monochromatic. Fix $\ell = 2$ or 3.*
Then one of the following holds.

(a) *For $u \in \Gamma_\ell$ and C a $\overset{2}{\sim}$-class of Γ_1, $u^1 \cap C$ is a proper subset of C, possibly empty.*

(b) *$m = 2$, $\ell = 3$, there is no monochromatic triangle of type 3, $\Gamma_3 \cong K_2^1[K_\infty^2]$, and for $u \in \Gamma_3$, $u^1 \cap \Gamma_1$ is a $\overset{2}{\sim}$-class of Γ_1.*

PROOF. We suppose that for $u \in \Gamma_\ell$, u^1 contains some $\overset{2}{\sim}$-class of Γ_1, and we show that we arrive at the specified exceptional case.

CLAIM 1. *For $u \in \Gamma_\ell$, $u^1 \cap \Gamma_1$ is a $\overset{2}{\sim}$-class of Γ_1.*

Let $A(u)$ be the finite set

$$\{C \in (\Gamma_1/\overset{2}{\sim}) \mid C \subseteq u^1\}$$

and set $k = |A(u)|$. We have assumed $k \geq 1$. Our claim is that $k = 1$.

Here we will use the fact that there are only three non-trivial 2-types in Γ.

Taking $u, v \in \Gamma_\ell$ distinct with $u \overset{1}{\perp} v$, we have $A(u) \cap A(v) = \emptyset$, and thus

$$k \leq m/2.$$

Suppose toward a contradiction that

$$k \geq 2.$$

By homogeneity, any k-set of $\overset{2}{\sim}$-classes in Γ_1 will occur as $A(u)$ for some $u \in \Gamma_\ell$.

For $u, v \in \Gamma_\ell$, the quantity

$$|A(u) \cap A(v)|$$

is determined by the type of (u, v) (which determines the type of u, v over the basepoint).

There are at least k possible values for $|A(u) \cap A(v)|$, namely $0, 1, \ldots, k-1$, and only three non-trivial 2-types.

Furthermore the case $|A(u) \cap A(v)| = k - 1$ must correspond to at least two distinct 2-types, as we can choose triples (c_1, c_2, c_3) and (c_1', c_2', c_3') in Γ_ℓ so that the the corresponding sets $A_i = A(c_i)$ and $A_i' = A(c_i')$ satisfy

$$|A_i \cap A_j| = |A_i' \cap A_j'| = k - 1 \ (i, j \text{ distinct});$$

$$|A_1 \cup A_2 \cup A_3| = k + 1; \qquad |A_1' \cup A_2' \cup A_3'| = k + 2.$$

Thus the triples (c_1, c_2, c_3) and (c_1', c_2', c_3') realize distinct types, and consequently some pairs (c_i, c_j) and (c_i', c_j') also realize distinct types, by homogeneity.

So at this point we have accounted for at least $k + 1$ non-trivial 2-types realized in Γ_ℓ, and thus we have only the case

$$k = 2$$

with all three 2-types accounted for.

In particular the possibility $A(u) = A(v)$ with $u, v \in \Gamma_\ell$ distinct does not occur. That is, $A(u)$ determines u.

But then the type of the pair $(A(u), A(v))$ determines the type of (u, v). But the type of the pair $(A(u), A(v))$ is determined by $|A(u) \cap A(v)|$, and as we have seen, this does not determine the type of (u, v), when $|A(u) \cap A(v)| = k - 1$.

This is a contradiction, and thus $k = 1$, as claimed.

Now for $u \in \Gamma_\ell$, let $C(u)$ denote the $\overset{2}{\sim}$-class $u^1 \cap \Gamma_1$ of Γ_1.

CLAIM 2. *The relation $\sim$ on Γ_ℓ defined by*

$$C(x) = C(y)$$

is non-trivial, with infinite equivalence classes, and in particular Γ_ℓ is imprimitive.

Let v_* be the chosen basepoint for Γ, and let $a \in \Gamma_1$.

We have

$$v_* \in a^1 \cong K_m^3[K_\infty^2].$$

Hence a^1 meets each of Γ_2 and Γ_3 in an infinite subset. Thus $a^1 \cap \Gamma_\ell$ is infinite.

But for $u \in a^1 \cap \Gamma_\ell$, we have $C(u) = a/\overset{2}{\sim}$. This proves that the $\sim$-classes in Γ_ℓ are infinite. As every $\overset{2}{\sim}$-class of Γ_1 occurs as $C(u)$ for some $u \in \Gamma_\ell$, the equivalence relation $\sim$ is non-trivial.

This proves the claim.

Now we know that Γ_ℓ realizes at least two 2-types, though it may possibly omit the type $\overset{\ell}{-}$. Also, the 2-type $\overset{1}{-}$ is realized in Γ_ℓ, and any realization involves a pair of points in distinct $\sim$-classes.

CLAIM 3. *The relation $\sim$ is a congruence on Γ_ℓ.*

We suppose the contrary. Then $\sim$ must be $\overset{i}{\sim}$ for some $i \neq 1$, and the other two 2-types are realized between any two $\sim$-classes of Γ_ℓ.

Let C_1, C_2 be distinct $\overset{2}{\sim}$-classes of Γ_1, and C_1^*, C_2^* be the corresponding $\sim$-classes of Γ_ℓ. Take $v \in C_1^*$, $u \in C_2^*$ with $u \overset{1}{-} v$. Then

$$C(u) \cup \{v\} \subseteq u^1.$$

In view of the structure of u^1, for some $i = 2$ or 3 we have

$$v \overset{i}{-} C(u).$$

If we vary u, but continue to require $u \overset{1}{-} v$, then $C(u)$ varies over $\Gamma_1 \setminus C(v)$, and i is fixed. Thus

$$v \overset{i}{-} \Gamma_1 \setminus C(v).$$

Thus v realizes only the types 1 and i over Γ_1. By homogeneity the same applies to all $v \in \Gamma_\ell$, with the same value of i. But all three 2-types are realized between Γ_1 and Γ_ℓ, so we arrive at a contradiction. The claim follows.

CLAIM 4. $m = 2$ and $\Gamma_\ell \cong K_2^1[C^*]$ with C^* a $\sim$-class.

We have a bijection between $\Gamma_1/\overset{2}{\sim}$ and $\Gamma_\ell/\sim$ definable over the basepoint. As $\Gamma_1/\overset{2}{\sim}$ realizes a unique non-trivial 2-type, the same applies to $\Gamma_\ell/\sim$. On the other hand the 2-type 1 is realized between $\sim$-classes, so we find

$$\Gamma_\ell \cong K_m^1[C^*]$$

with C^* a $\sim$-class.

Since there is no monochromatic triangle of type 1, we find $m = 2$.

CLAIM 5. The relation $\sim$ is the restriction of $\overset{2}{\sim}$ to Γ_ℓ, $\ell = 3$, and there is no monochromatic triangle of type 3 in Γ.

For $a \in \Gamma_1$,

$$a^1 \cap \Gamma_\ell$$

is a $\sim$-class of Γ_ℓ.

On the other hand a^1 contains the basepoint v_* and meets both Γ_2 and Γ_3. Since $a^1 \cong K_2^3[K_\infty^2]$ it follows that the $\overset{2}{\sim}$-class of a^1 containing v_* is $a^1 \cap (\Gamma_2 \cup \{v_*\})$ and the other $\overset{2}{\sim}$-class of a^1 is $a^1 \cap \Gamma_3$. In particular the only non-trivial 2-type realized in $a^1 \cap \Gamma_\ell$ is type 2, so $\sim$ is $\overset{2}{\sim}$.

Thus $\Gamma_\ell \cong K_2^1[K_\infty^2]$ omits type 3, and hence

$$\ell = 3$$

and Γ contains no monochromatic triangle of type 3.

The claim follows.

Now the lemma follows from Claims 1, 4, and 5.   □

Now, as promised, we eliminate the exceptional case which arose in the previous lemma.

LEMMA 19.3.9. Suppose that Γ is a locally degenerate imprimitive homogeneous 3-multi-graph with $\Gamma_1 \cong K_m^3[K_\infty^2]$ and $m < \infty$, and that Γ realizes all triangle types which are not monochromatic. Fix $\ell = 2$ or 3.

Then for $u \in \Gamma_\ell$ and C a $\overset{2}{\sim}$-class of Γ_1, $u^1 \cap C$ is a proper subset of C, possibly empty.

PROOF. We assume the contrary. By Lemma 19.3.8 we then have the following conditions.

– $m = 2$ and $\ell = 3$.
– There is no monochromatic triangle of type 3 in Γ.
– $\Gamma_3 \cong K_2^1[K_\infty^2]$.
– For $u \in \Gamma_3$, $u^1 \cap \Gamma_1$ is a $\overset{2}{\sim}$-class of Γ_1.

In particular restriction of the relation $\overset{2}{\sim}$ to either Γ_1 or Γ_3 is an equivalence relation.

CLAIM 1. *For $u \in \Gamma_1$, $u^1 \cap \Gamma_3$ contains a $\overset{2}{\sim}$-class of Γ_3.*

We consider $v \in u^1 \cap \Gamma_3$ and the $\overset{2}{\sim}$-class C of v in Γ_3. The equivalence relation defined on Γ_3 by "$x^1 \cap \Gamma_1 = y^1 \cap \Gamma_1$" has two classes, so all elements of C lie in the same class. The claim follows.

CLAIM 2. *For $v \in \Gamma_2$, $v^3 \cap \Gamma_3$ contains a $\overset{2}{\sim}$-class of Γ_3.*

We take $u \in v^1 \cap \Gamma_1$ and let $C^* = u^1 \cap \Gamma_3$. Then u^1 contains

$$C^* \cup \{v, v_*\}$$

where v_* is the chosen basepoint.

Here $v_* \overset{2}{-} v$ and $v_* \overset{3}{-} C^*$, so by virtue of the structure of u^1 we have $v \overset{3}{-} C^*$, and the claim follows.

Now consider the structure Γ' derived from Γ by interchanging the 2-types $1, 3$. This satisfies the hypotheses of the previous lemma. Furthermore, $\Gamma'_2 = \Gamma_2$, and Claim 2 may be expressed as follows, in terms of Γ': for $v \in \Gamma'_2$, $v^1 \cap \Gamma'_1$ contains a $\overset{2}{\sim}$-class of Γ'_1. Taking $\ell = 2$, this contradicts the previous lemma. This contradiction completes the proof.   □

LEMMA 19.3.10. *Suppose that Γ is a locally degenerate imprimitive homogeneous 3-multi-graph with $\Gamma_1 \cong K_m^3[K_\infty^2]$ and $m < \infty$, and that Γ realizes all triangle types which are not monochromatic. Then for $u \in \Gamma_2$ we have*

$$|u^1 \cap \Gamma_1| > 1.$$

PROOF. Supposing the contrary, we have a function $f : \Gamma_2 \to \Gamma_1$ defined by

$$u \overset{1}{-} f(u).$$

If this function is a bijection, then since Γ_1 realizes only two 2-types, the same applies to Γ_2. But as Γ_1 contains a monochromatic triangle of type 2, this is already a contradiction. So the following three cases are all witnessed by pairs in Γ_2.

$$f(u) = f(v), u \neq v,$$

$$f(u) \overset{2}{-} f(v),$$

$$f(u) \overset{3}{-} f(v).$$

Therefore the three 2-types realized in Γ_2 correspond to these three relations, in some order. So the relation $f(u) = f(v)$ is $\overset{k}{\sim}$ for some $k \in \{1, 2, 3\}$, and then since f gives a bijection of $\Gamma_2/\overset{i}{\sim}$ with $\Gamma_1 \cong K_m^i[K_\infty^2]$, Γ_2 has the structure

$$K_m^i[K_\infty^j[K_r^k]]$$

for some r, $2 \leq r \leq \infty$, where i, j, k are $1, 2, 3$ in some order.

As there are no monochromatic triangles of type 1, but there are triangles of type $(1, 1, k)$, it follows that k is not 1. Similarly j is not 1. So $i = 1$.

For $a \in \Gamma_2$, it follows that $a^1 \cap \Gamma_2 \cong K^1_{m-1}[K^j_\infty[K^k_r]]$. In particular a^1 contains a copy of $K^j_\infty[K^k_r]$. But this is impossible. $\qquad\square$

Now we continue with the idea that relevant finite numerical parameters should be 0 or 1, up to a change of notation (so really, the expected values are 0, 1, $m - 1$, or m, in our context). The next lemma is in this vein.

LEMMA 19.3.11. *Suppose that* Γ *is a locally degenerate imprimitive homogeneous 3-multi-graph with* $\Gamma_1 \cong K^3_m[K^2_\infty]$ *and* $m < \infty$, *and that* Γ *realizes all triangle types which are not monochromatic. Then one of the following conditions holds.*

(a) *For* $u \in \Gamma_2$, u^1 *meets each* $\overset{2}{\sim}$*-class of* Γ_1.

(b) *For* $u \in \Gamma_2$, u^1 *meets just one* $\overset{2}{\sim}$*-class of* Γ_1.

PROOF. Suppose toward a contradiction that for $u \in \Gamma_2$, u^1 meets m' $\overset{2}{\sim}$-classes of Γ_1, with

$$2 \le m' < m.$$

CLAIM 1. *Let* $\mathcal{C}$ *be a family of* $\overset{2}{\sim}$*-classes in* Γ_1 *with* $|\mathcal{C}| = m'$. *Let* A, A' *be two disjoint sets of representatives for* $\mathcal{C}$ *in* Γ_1. *Then there is some* $u \in \Gamma_2$ *with*

$$A \subseteq u^1$$
$$A' \cap u^1 = \emptyset.$$

By homogeneity, it suffices to show that a configuration isomorphic to $A \cup A' \cup \{u\}$ occurs with $A \cup A' \subseteq \Gamma_1$ and $u \in \Gamma_2$.

We start with $u \in \Gamma_2$, consider the associated family $\mathcal{C}_u$ of $\overset{2}{\sim}$-classes in Γ_1 which meet u^1, take A_u as a set of representatives for $\mathcal{C}_u$ lying in u^1, and then, since u^1 does not contain any of the $\overset{2}{\sim}$-classes of Γ_1, take A'_u as a set of representatives for $\mathcal{C}_u$ disjoint from u^1. Then $A_u \cup A'_u \cup \{u\}$ has the required form. This proves the claim.

Now fix two distinct families $\mathcal{C}_1, \mathcal{C}_2$ of $\overset{2}{\sim}$-classes in Γ_1, each consisting of m' classes, and with at least one class in common. Let A_1, A'_1, A_2, A'_2 be sets of representatives in Γ_1 for the family $\mathcal{C}_1$, and A_3, A'_3 sets of representatives in Γ_1 for the family $\mathcal{C}_2$, satisfying the following conditions.

(a) A_i, A'_i are disjoint for $i = 1, 2, 3$.

(b) $A_1 \cap A_2 \cap A_3$ is nonempty.

(c) A_2 meets A'_1.

Take elements u_1, u_2, u_3 in Γ_2 with

$$A_i \subseteq u^1_i \cap \Gamma_1;$$
$$A'_i \cap u^1_i = \emptyset.$$

Take $a \in A_1 \cap A_2 \cap A_3$. Then u_1, u_2, u_3 are in a^1 and so is the chosen basepoint v_*. As $v_* \overset{2}{=} u_1, u_2, u_3$ and $\overset{2}{\sim}$ is an equivalence relation on a^1, the three elements u_1, u_2, u_3 are all related by the type 2. Thus the configurations induced on v_*, u_1, u_2 and v_*, u_1, u_3 are isomorphic.

However the 3-types realized are different: u_1^1 and u_2^1 meet the same $\overset{2}{\sim}$-classes of Γ_1, while u_1^1 and u_3^1 do not meet the same $\overset{2}{\sim}$-classes of Γ_1. This is a contradiction. $\qquad\square$

Now we are more or less within striking distance of Lemma 19.3.19: for $u \in \Gamma_2$ and any non-trivial 2-type p, u^p meets all $\overset{2}{\sim}$-classes of Γ_1. But we will still need to eliminate a certain number of concrete alternatives. The first order of business is to understand where in fact the previous lemma leaves us. Some further notation will be useful, which is explained by the following lemma.

LEMMA 19.3.12. *There is a partition Π of the non-trivial 2-types $\{1, 2, 3\}$ such that for $u \in \Gamma_2$ and C a $\overset{2}{\sim}$-class of Γ_1, the set of 2-types realized by pairs (c, u) with $c \in C$ belongs to Π.*

PROOF. For $u \in \Gamma_2$ and C a $\overset{2}{\sim}$-class of Γ_1, let

$$u(C) = \{p \mid u^p \text{ meets } C\}.$$

CLAIM 1. *With $u \in \Gamma_2$ fixed, the sets $u(C)$ partition the set of nontrivial 2-types.*

The sets $u(C)$ are certainly non-empty.

Every non-trivial 2-type is realized by a pair u, v with $u \in \Gamma_2$, $v \in \Gamma_1$, and by homogeneity the same applies with $u \in \Gamma_2$ fixed. So it suffices to show that the sets $u(C)$ are pairwise disjoint or equal.

We suppose u^p meets the $\overset{2}{\sim}$-classes C_1, C_2 of Γ_1, and we take $c_i \in C_i$ with $u \overset{p}{=} c_i$ for $i = 1, 2$. Then u, c_1 and u, c_2 realize the same type over v_* and thus the set of types q for which u^q meets $c_1/\overset{2}{\sim}$ or $c_2/\overset{2}{\sim}$ are the same. This proves the claim.

Thus each $u \in \Gamma_2$ defines a corresponding partition Π_u of $\{1, 2, 3\}$. By homogeneity, the partition Π_u is independent of the choice of $u \in \Gamma_2$. $\qquad\square$

NOTATION 19.3.13. Π denotes the partition of the non-trivial 2-types $\{1, 2, 3\}$ given by Lemma 19.3.12. This will be called the (Γ_1, Γ_2)-*type partition*. Note that this terminology is not at all symmetrical.

For $u \in \Gamma_2$ and C a $\overset{2}{\sim}$-class of Γ_1, we set

$$u(C) = \{p \mid u^p \text{ meets } C\}$$

(a set belonging to Π).

Using this notation we can put Lemma 19.3.11 in a much clearer form.

LEMMA 19.3.14. *Suppose that* Γ *is a locally degenerate imprimitive homogeneous 3-multi-graph with* $\Gamma_1 \cong K_m^3[K_\infty^2]$ *and* $m < \infty$, *and that* Γ *realizes all triangle types which are not monochromatic. Let* Π *be the* (Γ_1, Γ_2)-*type partition. Then one of the following holds.*

1. *The partition* Π *is trivial: for all 2-types* p *and all* $u \in \Gamma_2$, u^p *meets each* $\overset{2}{\sim}$-*class of* Γ_1.

2. *The partition* Π *has the form* $(1i|j)$ *where* $\{i, j\} = \{2, 3\}$. *In this case, for all* $u \in \Gamma_2$ *we have the following.*

   – *The sets* $u^1 \cap \Gamma_1$ *and* $u^i \cap \Gamma_1$ *partition one* $\overset{2}{\sim}$-*class of* Γ_1,

   – *The set* $u^j \cap \Gamma_1$ *is the union of the other* $(m - 1)$ $\overset{2}{\sim}$-*classes.*

PROOF. If u^1 meets every $\overset{2}{\sim}$-class of Γ_1 then all of the sets $u(C)$ coincide and the partition Π is trivial. This is the first of our two possibilities.

Otherwise, the partition Π has at least two parts, and by Lemma 19.3.11, for fixed $u \in \Gamma_2$, the part containing the 2-type 1 is associated with a unique $\overset{2}{\sim}$-class of Γ_1.

On the other hand, by Lemma 19.3.9 the class of the 2-type 1 in the partition Π has at least two elements, and thus Π has the form $(12|3)$ or $(13|2)$.

The lemma follows. $\square$

LEMMA 19.3.15. *Suppose that* Γ *is a locally degenerate imprimitive homogeneous 3-multi-graph with* $\Gamma_1 \cong K_m^3[K_\infty^2]$ *and* $m < \infty$, *and that* Γ *realizes all triangle types which are not monochromatic. Suppose that the* (Γ_1, Γ_2)-*type partition is non-trivial. Let* $\sim$ *be the equivalence relation on* Γ_2 *defined by*

$$(x^2 \cap \Gamma_1)/\overset{2}{\sim} = (y^2 \cap \Gamma_1)/\overset{2}{\sim}.$$

Then $\sim$ *is either* $\overset{1,2}{\sim}$ *or* $\overset{2,3}{\sim}$ *(that is, the reflexive extension of the union of types 1 and 2 or types 2 and 3, respectively), and the following hold.*

(a) *If* $\sim$ *is* $\overset{1,2}{\sim}$ *then*

   – *for* $u_1, u_2 \in \Gamma_1$ *satisfying* $u_1 \overset{2}{-} u_2$, *the sets* $u_1^2 \cap \Gamma_1$ *and* $u_2^2 \cap \Gamma_1$ *meet, and*

   – Γ_2 *contains a triangle of type* $(2, 2, 1)$.

(b) *If* $\sim$ *is* $\overset{2,3}{\sim}$ *then*

   – *for* $u_1, u_2 \in \Gamma_2$ *satisfying* $u_1 \overset{2}{-} u_2$, *the sets* $u_1^2 \cap \Gamma_1$ *and* $u_2^2 \cap \Gamma_1$ *are disjoint, and*

   – *the* (Γ_1, Γ_2)-*type partition is* $(12|3)$.

PROOF. By hypothesis Case 2 of Lemma 19.3.14 applies. Thus we have a definable map

$$\gamma : \Gamma_2 \to \Gamma_1/\overset{2}{\sim}$$

given by $\gamma(u) = (u^1 \cap \Gamma_1)/\overset{2}{\sim}$.

Furthermore $(u^2 \cap \Gamma_1)/\overset{2}{\sim}$ is either $\gamma(u)$ or $\Gamma_1 \setminus \gamma(u)$, so $\sim$ is the kernel of the map γ. By homogeneity γ is surjective, so

$$|\Gamma/\!\sim| = m.$$

Any non-trivial 2-type is either contained in the equivalence relation $\sim$ or transversal to it, on Γ_2.

As Γ contains an infinite 2-clique, so does Γ_2. As $|\Gamma_2/\!\sim|$ is finite, the relation $\overset{2}{\sim}$ is contained in $\sim$.

Furthermore some 2-type is not contained in $\sim$.

The rest of the analysis divides into two cases, each of which leads to one of the two alternatives in the statement of the lemma.

Case 1. For $u_1, u_2 \in \Gamma_2$ satisfying $u_1 \overset{2}{\text{---}} u_2$, the sets $u_1^2 \cap \Gamma_1$ and $u_2^2 \cap \Gamma_2$ meet.

Take $u_1, u_2 \in \Gamma_2$ with $u_1 \overset{2}{\text{---}} u_2$ and take $v \in u_1^2 \cap u_2^2 \cap \Gamma_1$. Then u_1^2 contains the triangle (u_2, v, v_*) with v_* the basepoint. This is a triangle of type $(2, 2, 1)$, so this triangle type is realized in Γ_2. As $\overset{2}{\sim}$ is contained in the equivalence relation $\sim$, the relation $\overset{1}{\sim}$ is also contained in $\sim$. Since some 2-type is not contained in $\sim$, we find that $\sim$ is $\overset{1,2}{\sim}$ on Γ_2.

Since we have arrived at the first of the two possibilities envisioned in the statement of the lemma, this disposes of the first case.

Case 2. For $u_1, u_2 \in \Gamma_2$ satisfying $u_1 \overset{2}{\text{---}} u_2$, the sets $u_1^2 \cap \Gamma_1$ and $u_2^2 \cap \Gamma_2$ are disjoint.

As $\overset{2}{\sim}$ is contained in $\sim$, u_1^2 and u_2^2 meet the same $\overset{2}{\sim}$-classes in Γ_1. If $u_1^2 \cap \Gamma_1$ and $u_2^2 \cap \Gamma_1$ are disjoint, it follows that these sets split the relevant classes non-trivially. Therefore the (Γ_1, Γ_2)-type partition Π must be

$$(12|3).$$

It remains to show that the equivalence relation $\sim$ is $\overset{2,3}{\sim}$.

Suppose first that for $u \in \Gamma_2$ we have
$(\star)$ $|u^2 \cap \Gamma_1| = 1$.

Consider the function $f : \Gamma_2 \to \Gamma_1$ defined by

$$x \overset{2}{\text{---}} f(x).$$

For $u \in \Gamma_2$ and $C = f(u)/\overset{2}{\sim}$ we have

$$u^1 \cap \Gamma_1 = C \setminus \{f(u)\}.$$

Thus for $u \sim v$ in Γ_2 the sets $u^1 \cap \Gamma_1$ and $v^1 \cap \Gamma_1$ intersect. In particular the type $\overset{1}{\text{---}}$ is disjoint from $\sim$.

If $\sim$ is $\overset{2}{\sim}$ then f is a bijection. As Γ_1 realizes two 2-types and Γ_2 realizes 3, this is a contradiction.

So $\sim$ must be $\overset{2,3}{\sim}$.

Now suppose that for $u \in \Gamma_2$ we have

$(\star)$ $|u^2 \cap \Gamma_1| > 1$.

Fix $a \in \Gamma_1$ and let $C = a/\overset{2}{\sim}$. We claim that the family

$$\mathcal{F}_a = \{u^2 \cap C \mid u \in a^2 \cap \Gamma_2\}$$

is infinite.

By assumption the sets in $\mathcal{F}_a$ do not reduce to a, so their union is an a-definable subset of C properly containing $\{a\}$. Therefore the union $\bigcup \mathcal{F}_a$ is C.

Thus if the sets in $\mathcal{F}_a$ are finite, there are certainly infinitely many of them.

If the sets in $\mathcal{F}_a$ are infinite, then by homogeneity any finite subset of C is contained in a set in $\mathcal{F}_a$, and as each set in $\mathcal{F}_a$ is a proper subset of C, it again follows that the family $\mathcal{F}_a$ is infinite.

In particular, for $a \in \Gamma_1$ the set $a^2 \cap \Gamma_2$ is infinite.

Consider the following three relations on Γ_2.

$$x \not\sim y;$$
$$x \sim y, x^2 \cap y^2 = \emptyset;$$
$$x \sim y, x^2 \cap y^2 \neq \emptyset, x \neq y.$$

Each of these three relations is realized in Γ_2, and hence they coincide in some order with the three non-trivial 2-types in Γ_2. By hypothesis, type 2 implies, and hence coincides with, the second of these relations.

For distinct elements of $a^2 \cap \Gamma_2$ the third relation holds, so $a^2 \cap \Gamma_2$ is monochromatic of the type associated with that relation; so this cannot be type 1. Therefore it is type 3, and $\overset{3}{\sim}$ is contained in $\sim$.

Thus again $\sim$ is $\overset{2,3}{\sim}$.

So the second case leads to our second alternative.

This completes the proof of all parts of the lemma. $\qquad\square$

For the present we will continue to use the symbol $\sim$ for the canonical equivalence relation on Γ_2, as long as we consider the configurations envisioned above with a non-trivial (Γ_1, Γ_2)-type partition. And when we switch basepoint, considering u^2 rather than Γ_2, we will use $\sim$ in a similar sense, without changing the notation.

LEMMA 19.3.16. *Suppose that Γ is a locally degenerate imprimitive homogeneous 3-multi-graph with $\Gamma_1 \cong K_m^3[K_\infty^2]$ and $m < \infty$, and that Γ realizes all triangle types which are not monochromatic. Suppose that the (Γ_1, Γ_2)-type partition is non-trivial. Then the (Γ_1, Γ_2)-type partition is $(12|3)$.*

PROOF. Suppose the contrary, so that by Lemma 19.3.14 the (Γ_1, Γ_2)-type partition is

$$(13|2).$$

Then by Lemma 19.3.15 we have the following conditions.

(a) The relation $\sim$ is $\overset{1,2}{\sim}$
(b) For $u_1, u_2 \in \Gamma_2$ satisfying $u_1 \overset{2}{\longrightarrow} u_2$, the sets $u_1^2 \cap \Gamma_1$ and $u_2^2 \cap \Gamma_2$ meet.
(c) Γ_2 contains a triangle of type $(2, 2, 1)$.

CLAIM 1. $m = 2$.

For $u \in \Gamma_2$ we have the following configuration inside u^2, with v_* the basepoint as usual.

$$v_* \sim u^2 \cap \Gamma_1.$$

Hence the type 3 is not realized in $u^2 \cap \Gamma_1$.

However the (Γ_1, Γ_2)-type partition Π is $(13|2)$, so $u^2 \cap \Gamma_1$ is a union of $(m-1)$ $\overset{2}{\sim}$-classes of Γ_1. Thus $m = 2$.

CLAIM 2. *For* $v \in \Gamma_3$, *the set* $v^3 \cap \Gamma_1$ *is a* $\overset{2}{\sim}$-*class of* Γ_1.

Take $u \in v^2 \cap \Gamma_2$ and let $C = u^2 \cap \Gamma_1$; note that since $m = 2$ and Π is $(13|2)$, this is a $\overset{2}{\sim}$-class in Γ_1.

In u^2 we have the following.

$$v_* \sim C; \qquad v_* \overset{3}{\longrightarrow}, v$$

so $v \overset{3}{\longrightarrow} C$. As $m = 2$ and Γ_1 is not contained in v^3 (the types 1, 2 are also realized by v over Γ_1) we find $v^3 \cap \Gamma_1 = C$.

This proves the claim.

Now we reach a contradiction. Take a class $C \in \Gamma_1 / \overset{2}{\sim}$ and $a, b \in C$. There is $u \in a^2$ with $b \overset{3}{\longrightarrow} u$.

Clearly $u \notin \Gamma_1$.

If $u \in \Gamma_2$ then as $a \in u^2$ we have $u^2 \cap \Gamma_1 = C$ and $b \in u^2$, a contradiction.

If $u \in \Gamma_3$ then as $b \in u^3$ we have $u^3 \cap \Gamma_1 = C$ and $a \in u^3$, a contradiction.

Thus u does not exist, and this contradiction proves the lemma. $\square$

We mentioned the next lemma earlier, as an application of Example 19.3.7.

LEMMA 19.3.17. *Let* Γ *be a binary homogeneous structure with two distinct 1-types over the empty set, with loci* A *and* B. *Suppose that* A *carries a non-trivial $\emptyset$-definable equivalence relation* $\sim$ *with finitely many classes, and with a unique type realized between any two distinct classes. Let* p *be a 2-type realized in* $B \times A$. *If* B *realizes at most 3 2-types, then for* $b \in B$, b^p *is not a transversal to* $\sim$ *in* A.

PROOF. We suppose the contrary. Let T be the set of transversals to $\sim$ in A and let

$$\tau : B \to T$$

be defined by $\tau(b) = b^p$.

Note that all transversals realize the same type in Γ. In particular τ is surjective.

There are at least two non-trivial binary relations defined on T, namely

$$t_1 \cap t_2 = \emptyset;$$
$$t_1 \cap t_2 \neq \emptyset, t_1 \neq t_2.$$

Define the equivalence relation $\approx$ on B by

$$\tau(x) = \tau(y).$$

As B realizes at most three 2-types, this is a congruence (possibly trivial).

So $B/\approx$ is again a binary homogeneous structure. But there is a definable bijection with T, which is not a binary homogeneous structure, as there are 4-types not determined by their restrictions to 2-types.

Namely, when there are just two $\approx$-classes then T is a kind of grid, as discussed in Example 19.3.7. The 4-types in question involve two pairs of grid segments in general position, parallel in one case and not in the other—for example, the 4-tuples

$$(0,0), (1,0), (2,1), (3,1)$$
$$(0,0), (1,0), (2,1), (2,2)$$

in $\{0,1,2,3\}^2$.

When there are more than 2 classes we arrive at the same situation by fixing $m-2$ elements of the transversals involved.

This completes the proof.   □

LEMMA 19.3.18. *Suppose that Γ is a locally degenerate imprimitive homogeneous 3-multi-graph with $\Gamma_1 \cong K_m^3[K_\infty^2]$ and $m < \infty$, and that Γ realizes all triangle types which are not monochromatic. Suppose that the (Γ_1, Γ_2)-type partition is non-trivial. Then the relation $\sim$ is $\overset{2,3}{\sim}$.*

PROOF. Supposing the contrary, by Lemma 19.3.15 we have the following conditions.

(a) The relation $\sim$ is $\overset{1,2}{\sim}$.
(b) For $u_1, u_2 \in \Gamma_2$ satisfying $u_1 \overset{2}{-} u_2$, the sets $u_1^2 \cap \Gamma_1$ and $u_2^2 \cap \Gamma_1$ meet.
(c) Γ_2 contains a triangle of type $(2,2,1)$.

CLAIM 1. *Let C be a $\sim$-class of Γ_2 and $v \in \Gamma_3$. Then*

1.1 v^p meets C for all 2-types p;
1.2 $|v^2 \cap C| = 1$.

It suffices to show this for one choice of v and C. Then the first clause applies to all such pairs and hence the second one does as well. So we may assume from the outset that v^2 meets C.

It will be convenient to use notation analogous to that of Notation 19.3.13; namely, we may consider the (Γ_2, Γ_3)-type partition encoding the combinations of types realized by elements of Γ_3 over $\sim$-classes of Γ_2.

Fix $a \in v^2 \cap C$. Then in a^2 we have the following.

$$v_* \overset{2}{-} a^2 \cap C;$$

$$v_* \overset{3}{-} v.$$

In view of the structure of Γ_2 this gives

$$v \overset{3}{-} a^2 \cap C.$$

So at least v^3 meets C.

If $v^1 \cap C = \emptyset$ then there is a (Γ_2, Γ_3)-partition of the non-trivial 2-types of the form $(1|23)$ and thus $v^1 \cap \Gamma_2$ is a union of $\sim$-classes. In particular v^1 realizes the type 1, which is a contradiction.

So v^1, v^2, and v^3 all meet C, and as we have seen, $a^2 \cap C \subseteq v^3$.

Now suppose $|v^2 \cap C| > 1$ and fix $b \in v^2 \cap C$, $b \neq a$. Since we have $a^2 \cap C \subseteq v^3$, we cannot have $a \overset{2}{-} b$, and since $a \sim b$ we must have

$$b \in a^1.$$

On the other hand, there is no triangle of type $(1, 1, 1)$ and hence

$$b \overset{2}{-} (a^1 \cap C) \setminus \{b\}.$$

As above, $b^2 \cap C \subseteq v^3$. But then $C \subseteq v^2 \cup v^3$, a contradiction. This proves the claim.

Now apply Lemma 19.3.17 to the structure (Γ_2, Γ_3) to get a contradiction. This proves the lemma. $\qquad\square$

LEMMA 19.3.19. *Suppose that* Γ *is a locally degenerate imprimitive homogeneous 3-multi-graph with* $\Gamma_1 \cong K_m^3[K_\infty^2]$ *and* $m < \infty$, *and that* Γ *realizes all triangle types which are not monochromatic. Then for all* $u \in \Gamma_2$ *and all non-trivial 2-types* p, u^p *meets each* $\overset{2}{\sim}$*-class of* Γ_1.

PROOF. In other words, we claim that the (Γ_1, Γ_2) partition of 2-types is trivial. By Lemmas 19.3.16 and 19.3.18 the only other possibility is that the (Γ_1, Γ_2)-type partition is

$$(12|3)$$

and that the canonical equivalence relation $\sim$ on Γ_2 is the relation

$$\overset{2,3}{\sim}.$$

Recall that Γ_2 has m $\sim$-classes.

Since the type $\overset{1}{\perp}$ holds between points of distinct $\sim$-classes in Γ_2, and there is no monochromatic triangle of type 1, we find

$$m = 2.$$

CLAIM 1. *For $a \in \Gamma_1$, the structures induced on $a^1 \cap \Gamma_2$ and $a^1 \cap \Gamma_3$ are 2-cliques.*

In a^1 we have the following.

$$v_* \overset{2}{-} a^1 \cap \Gamma_2;$$

$$v_* \overset{3}{-} a^1 \cap \Gamma_3,$$

so the claim follows from the structure of Γ_1, bearing in mind that $m = 2$.

CLAIM 2. *For $u \in \Gamma_2$ the structures induced on the sets*

$$u^2 \cap \Gamma_i \ (i = 1, 2, 3)$$

are 2-cliques.

This is known for the case of $u^2 \cap \Gamma_1$.

Take $a \in u^2 \cap \Gamma_1$. In u^2 we have the following.

$$v_* \overset{1}{-} a;$$

$$v_* \sim u^2 \cap (\Gamma_2 \cup \Gamma_3).$$

So $a \overset{1}{\perp} u^2 \cap (\Gamma_2 \cup \Gamma_3)$ and the previous claim applies.

This proves the claim.

So u^2 is a union of three 2-cliques and hence the same applies to Γ_2. But then some $\sim$-class must be a 2-clique, hence all are, which is a contradiction. $\square$

19.3.3. Proof that $m = \infty$. Having disposed of a large number of special cases under the assumption that m is finite, we will shortly be able to exclude that case entirely (Proposition 19.3.24).

LEMMA 19.3.20. *Suppose that Γ is a locally degenerate imprimitive homogeneous 3-multi-graph with $\Gamma_1 \cong K_m^3[K_\infty^2]$ and $m < \infty$, and that Γ realizes all triangle types which are not monochromatic. Then for $u \in \Gamma_2$, all non-trivial 2-types p, and all $\overset{2}{\sim}$-classes C of Γ_1, u^p meets C in an infinite set.*

PROOF. By Lemma 19.3.19 u^p meets every $\overset{2}{\sim}$-class of Γ_1.

Applying Lemma 19.3.17 to the structure (Γ_1, Γ_2) we find that

$$|u^p \cap C| > 1$$

for all $\overset{2}{\sim}$-classes C in Γ_1.

Fix a $\overset{2}{\sim}$-class C and suppose toward a contradiction that p is a 2-type for which the sets $u^p \cap C$ are finite for $u \in \Gamma_2$.

$$|u^p \cap C| = n \text{ with } 1 < n < \infty.$$

Let p, i, j be the types $1, 2, 3$ in some order. For $u, v \in \Gamma_2$ and $S \subseteq \{i, j\}$ let

$$C_u = u^p \cap C;$$

$$S_C(u, v) = \{q \in \{i, j\} \mid v^q \cap C_u \neq \emptyset\};$$

$$\nu_S(u, v) = |\{C \in (\Gamma_1 / \overset{2}{\sim}) \mid S_C(u, v) = S\}|.$$

The condition $S_C(u, v) = \emptyset$ means that $v^p \cap C = u^p \cap C$. But for any choice of values for ν_S with S varying over nonempty subsets of $\{i, j\}$, if their sum is m then there is some $v \in \Gamma_2$ for which $\nu_S(u, v)$ takes on the specified values.

There are at least 6 possible choices of such functions and hence at least 6 non-trivial 2-types realized in Γ_2, which is a contradiction. $\square$

LEMMA 19.3.21. *Suppose that* Γ *is a locally degenerate imprimitive homogeneous 3-multi-graph with* $\Gamma_1 \cong K_m^3[K_\infty^2]$ *and* $m < \infty$, *and that* Γ *realizes all triangle types which are not monochromatic. Then the following hold.*

1. *For* $v \in \Gamma_2$, *we have* $v^1 \cap \Gamma_2 \cong K_m^3[K_\infty^2]$, *and each* $\overset{2}{\sim}$*-class of* v^1 *meets* Γ_1.

2. $\overset{2}{\sim}$ *does not define an equivalence relation on* Γ_2.

PROOF.

CLAIM 1. *For* $v \in \Gamma_2$, *every* $\overset{2}{\sim}$*-class of* v^1 *meets* Γ_1.

By Lemma 19.3.20, $v^1 \cap \Gamma_1$ has m $\overset{2}{\sim}$-classes. As v^1 also has m $\overset{2}{\sim}$-classes, the claim follows.

CLAIM 2. *For* $v \in \Gamma_2$, *the* $\overset{2}{\sim}$*-classes of* $v^1 \cap \Gamma_2$ *are infinite.*

Take $u \in \Gamma_2$, and C a $\overset{2}{\sim}$-class of Γ_1. It suffices to verify the claim with u^2 in place of Γ_2.

By Lemma 19.3.20, the set $u^2 \cap C$ is infinite. Let v_* be the chosen basepoint of Γ. Then in u^2, we have $v_* \overset{1}{\longrightarrow} u^2 \cap C$, so $v_*^1 \cap u^2$ has infinite $\overset{2}{\sim}$-classes.

The claim follows.

CLAIM 3. *For* $v \in \Gamma_2$, $v^1 \cap \Gamma_2 \cong K_m^3[K_2^\infty]$.

As the $\overset{2}{\sim}$-classes of $v^1 \cap \Gamma_2$ are infinite, it suffices to show that there are m of them.

Take $u \in v^1 \cap \Gamma_2$. Then the $\overset{2}{\sim}$-class of u in v^1 meets a unique $\overset{2}{\sim}$-class $C(v, u)$ of Γ_1. It follows that every $\overset{2}{\sim}$-class of Γ_1 occurs as $C(v, u)$ for some $u \in v^1 \cap \Gamma_2$ and the claim follows.

CLAIM 4. *The relation* $\overset{2}{\sim}$ *is not an equivalence relation on* Γ_2.

Take $u \in \Gamma_2$ and take distinct v_1, v_2 in a $\overset{2}{\sim}$-class of $u^1 \cap \Gamma_2$. Then take $c \in \Gamma_1$ with $v_1, v_2 \overset{2}{\longrightarrow} c$.

We switch to the point of view with v_1 as basepoint: that is, we consider v_1^2 in place of Γ_2. In v_1^2 we have

$$v_2 \overset{2}{\longrightarrow} v_*, c.$$

Since $v_* \overset{1}{\perp} c$, the relation $\overset{2}{\sim}$ is not an equivalence relation on v_1^2. Thus the same applies to Γ_2. The claim follows.

These claims prove the lemma.   □

LEMMA 19.3.22. *Suppose that Γ is a locally degenerate imprimitive homogeneous 3-multi-graph with $\Gamma_1 \cong K_m^3[K_\infty^2]$, $m < \infty$, and that Γ realizes all triangle types which are not monochromatic. Let $v_1, v_2 \in \Gamma_2$ with $v_1 \overset{2}{\perp} v_2$. Then the following hold.*

1. *For any non-trivial 2-type p, we have $v_1^p \cap \Gamma_1 \neq v_2^p \cap \Gamma_1$.*

2. *For any $\overset{2}{\sim}$-class C in Γ_1, there is a 2-type p for which $v_1^p \cap C \neq v_2^p \cap C$.*

PROOF.

CLAIM 1. *For each $\overset{2}{\sim}$-class C of Γ_1 and each non-trivial 2-type p, the family of sets*

$$\{u^p \cap C \mid u \in \Gamma_2\}$$

is infinite.

As $u^p \cap C$ is infinite, every finite subset of C is contained in one of the sets in this family. As these sets are proper subsets of C the claim follows.

Ad 1. Consider the equivalence relation $\approx$ defined on Γ_2 by

$$x^p \cap \Gamma_1 = y^p \cap \Gamma_1.$$

Suppose toward a contradiction that $\overset{2}{\sim}$ is contained in $\approx$. By Lemma 19.3.21, the relation $\overset{2}{\sim}$ is not itself an equivalence relation on Γ_2. Thus there is only one 2-type transversal to $\approx$ in Γ_2.

By Claim 1 the quotient $\Gamma_2/\approx$ is infinite, so this transversal 2-type is not the type 1. Thus for $u \overset{1}{\perp} v$ in Γ_2 we also have $u^p \cap \Gamma_1 = v^p \cap \Gamma_1$.

In particular p is not the type 1. Now take two $\overset{2}{\sim}$-classes C_1, C_2 in Γ_1 with $u^1 \cap C_1$ in the same $\overset{2}{\sim}$-class of u^1 as v.

$$u^1 \cap (C_1 \cup C_2) \overset{1}{\underset{\Gamma_1}{\perp}} u \overset{1}{\underset{\Gamma_2}{\perp}} v$$

In view of the structure of u^1, the vertex v realizes the type 2 over $u^1 \cap C_1$ and realizes the type 3 over $u^1 \cap C_2$. So the type p cannot be 2 or 3 either, and we have a contradiction.

Ad 2. Suppose toward a contradiction that C is a $\overset{2}{\sim}$-class in Γ_1 such that $v_1^p \cap C = v_2^p \cap C$ for all non-trivial 2-types p.

Take a vertex $u \in \Gamma$ with $u \overset{1}{\perp} v_1, v_2$ and consider an arbitrary $\overset{2}{\sim}$-class C' of Γ_1. As v_1, v_2 are in the same $\overset{2}{\sim}$-class of u^1, and are not in C', they realize a single type p over the vertices of $u^1 \cap C'$.

By Lemma 19.3.19 and our current hypothesis, the vertices v_1, v_2 also realize the type p over some point a of C. Since v_1, v_2 induce the same partition of $C = a/\overset{2}{\sim}$, by homogeneity v_1, v_2 induce the same partition of $(u^1 \cap C')/\overset{2}{\sim} = C'$. As C' is arbitrary this contradicts (1). The claim follows. This proves the lemma. $\square$

LEMMA 19.3.23. *Suppose that Γ is a locally degenerate imprimitive homogeneous 3-multi-graph with $\Gamma_1 \cong K_m^3[K_\infty^2]$, $m < \infty$, and that Γ realizes all triangle types which are not monochromatic. Let $u, v \in \Gamma_2$ with $u \overset{1}{\_} v$. Then the following hold.*

1. *For each $\overset{2}{\sim}$-class C of Γ_1, there is a non-trivial 2-type p for which the set $u^p \cap C$ is not v-definable in C.*

2. *For each non-trivial 2-type p, there is a $\overset{2}{\sim}$-class C of Γ_1 for which the set u^p is not v-definable.*

PROOF. *Ad* 1. Fix the $\overset{2}{\sim}$-class C. Suppose toward a contradiction that $u^p \cap C$ is v-definable for all 2-types p. Then the sets $u^p \cap C$ coincide with the sets $v^p \cap C$ up to a permutation.

By Lemma 19.3.21, $u^1 \cap \Gamma_2$ has non-trivial $\overset{2}{\sim}$-classes.

If v_1, v_2 are in the same $\overset{2}{\sim}$-class of $u^1 \cap \Gamma_2$, then they realize the same type over $u^1 \cap C$. and hence the relevant permutation of the sets $v_1^p \cap C$ or $v_2^p \cap C$ is the same. That is, $v_1^p \cap C = v_2^p \cap C$ for all 2-types p. This contradicts the previous lemma. Point (1) follows.

Ad 2. Let p be a non-trivial 2-type and let C be any $\overset{2}{\sim}$-class in Γ_1.

Suppose that $u^p \cap C$ is v-definable in C. Then for u_1, u_2 in the $\overset{2}{\sim}$-class of u in $v^1 \cap \Gamma_2$, since u_1 and u_2 realize the same type over the points of $v^1 \cap C$, we have $u_1^p \cap C = u_2^p \cap C = u^p \cap C$. By Lemma 19.3.22 this cannot hold for every $\overset{2}{\sim}$-class C of Γ_1. Point (2) follows. $\square$

After these lengthy preliminaries we can finally exclude the case in which m is finite.

PROPOSITION 19.3.24. *Suppose that Γ is a locally degenerate imprimitive homogeneous 3-multi-graph with $\Gamma_1 \cong K_m^3[K_\infty^2]$, and that Γ realizes all triangle types which are not monochromatic. Then $m = \infty$.*

PROOF. We fix $u, v \in \Gamma_2$ with $u \overset{1}{\_} v$. We choose two $\overset{2}{\sim}$-classes C_1, C_2 of Γ_1 so that $u^1 \cap C_1$ is in the same $\overset{2}{\sim}$-class of u^1 as v.

We study the partitions into three labeled parts induced on C_1 and on C_2 by u and v. This will lead to a configuration contradicting the previous lemma. To begin with, we have the following.

$$v \overset{2}{\_} u^1 \cap C_1; \qquad\qquad v \overset{3}{\_} u^1 \cap C_2. \qquad (19.1)$$

We may represent this by the following chart,where the partitions of C_1, C_2 afforded by u are indicated by numbers within the boxes representing C_1, C_2

and the partitions associated with v are indicated by numbers outside the boxes, when the indicated intersections are known to be non-empty. We add brackets when any entries not shown are known to correspond to empty intersections.

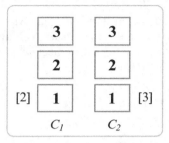

Maintaining a chart of this type may be helpful in following the analysis below.

An essential point is that the information concerning C_1 and C_2 is complementary in the following sense.

CLAIM 1. *No 1-type over the pair u, v is realized in both C_1 and C_2: that is, any set of the form $u^p \cap v^q$ meets at most one of the classes C_1, C_2.*

Otherwise, we could find an automorphism taking u, v, C_1 to u, v, C_2, and fixing the basepoint, contradicting the relations of $u^1 \cap C_1$ and $u_1 \cap C_2$ to v. This proves the claim.

CLAIM 2. *If $u^i \cap C_1$ meets $v^i \cap C_1$ for some i, then modulo the basepoint C_1 realizes the same type over u, v as over v, u. In other words, there is an automorphism fixing the basepoint and taking (C_1, u, v) to (C_1, v, u).*

In particular we then have

$$v^1 \cap C_1 \subseteq u^2 \cap C_1.$$

Indeed, if we have some $c \in C_1$ realizing the same type over u and v, then we can find an automorphism fixing the basepoint and taking cuv to cvu, hence taking (C_1, u, v) to (C_1, v, u). Then from $u^1 \cap C \subseteq v^2$ we deduce $v^1 \cap C \subseteq u^2$. This proves the claim

As we accumulate more information about C_1 this argument will have further consequences. It applies a little more generally: if both $u^i \cap v^j \cap C_1$ and $u^j \cap v^i \cap C_1$ are non-empty then again C_1, u, v and C_1, v, u realize the same type.

CLAIM 3. $v^1 \cap C_1 \neq u^3 \cap C_1$

If $v^1 \cap C_1 = u^3 \cap C_1$ then by Claim 2 $v^2 \cap C_1$ is disjoint from $u^2 \cap C_1$, and hence $v^2 \cap C_1$ coincides with $u^1 \cap C_1$, leaving $v^3 \cap C_1$ to be $u^2 \cap C_1$. At this point each of $u^1 \cap C_1$, $u^2 \cap C_1$, and $u^3 \cap C_1$ is v-definable, contradicting Lemma 19.3.23, part (1).

This proves the claim.

Claim 4. v^1 *meets* $u^2 \cap C_1$.

Suppose the contrary. Then

$$v^1 \cap C_1 \subseteq u^3 \cap C_1,$$

and by the previous claim this inclusion is proper. Also, under this hypothesis claim 2 still applies and in particular the sets $u^3 \cap C_1$ and $v^3 \cap C_1$ are disjoint.

Thus $v^3 \cap C_1 \subseteq u^2 \cap C_1$ and v^2 must meet $u^3 \cap C_1$. This situation is summarized in the following table.

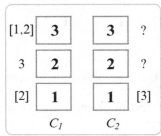

[1,2]	**3**	**3**	?
3	**2**	**2**	?
[2]	**1**	**1**	[3]
	C_1	C_2	

We also have $v^3 \cap C_1 = u^2 \cap C_1$, since we assume v_1 disjoint from $u^2 \cap C_1$ and Claim 2 gives v^2 disjoint from $u^2 \cap C_1$. So the left side of the table is completely precise.

Now by our complementarity principle neither v^1 nor v^2 meets $u^3 \cap C_2$, and hence

$$u^3 \cap C_2 \subseteq v^3.$$

Similarly, v^3 does not meet $u^2 \cap C_2$. Thus $v^3 \cap C_2 = (u^1 \cup u^3) \cap C_2$ is u-definable in C_2.

But $v^3 \cap C_1 = u^2 \cap C_1$ is also u-definable, and we arrive at a contradiction to Lemma 19.3.23, part (2), since all $\overset{2}{\sim}$-classes in Γ_1 have the same type over u, v as C_1 or C_2. This proves the claim.

Taking stock, at this point our reference chart looks as follows.

	3	**3**	
1	**2**	**2**	
[2]	**1**	**1**	[3]
	C_1	C_2	

Since $u^1 \cap v^2$ and $u^2 \cap v^1$ both meet C_1, arguing as in the proof of Claim 2 we now have the following symmetry principle.

Claim 5. (C_1, u, v) *and* (C_1, v, u) *have the same type over the basepoint.*

CLAIM 6. v^3 *meets* $u^2 \cap C_1$.

We suppose the contrary. Then

$$v^3 \cap C_1 \subseteq u^3.$$

In particular Claim 2 applies and as we have $u^1 \cap C_1 \subseteq v^2$ and $v^3 \cap C_1 \subseteq u^3$ we deduce

$$v^1 \cap C_1 \subseteq u^2; \qquad\qquad v^3 \cap C_1 = u^3 \cap C_1.$$

We claim that v^2 meets $u^2 \cap C_1$. Otherwise, we have $v^2 \cap C_1 = u^1 \cap C_1$ and then $v^1 \cap C_1$ must be $u^2 \cap C_1$ and the partition of C_1 induced by v is u-definable, contradicting Lemma 19.3.23, part (2).

So both v^1 and v^2 meet $u^2 \cap C_1$ and therefore neither meets $u^2 \cap C_2$.

$$u^2 \cap C_2 \subseteq v^3.$$

As $v^3 \cap C_1 = u^3 \cap C_1$, v^3 does not meet $u^3 \cap C_2$ and thus $v^3 \cap C_2 = (u^1 \cup u^2) \cap C_2$. Thus $v^3 \cap C_1$, $v^3 \cap C_2$ are both u-definable, contradicting Lemma 19.3.23, part (2).

This proves the claim.

CLAIM 7. $v^2 \cap C_2 = u^2 \cap C_2$.

As C_1, u, v and C_1, v, u realize the same type we deduce from the previous claim that v^2 meets $u^3 \cap C_1$, and thus v^2 does not meet $u^3 \cap C_2$. Thus

$$v^2 \cap C_2 \subseteq u^2.$$

As v^1, v^3 meet $u^2 \cap C_1$, they do not meet $u^2 \cap C_2$ and thus

$$u^2 \cap C_2 \subseteq v^2.$$

This proves the claim.

From this, it follows also that

$$v^1 \cap C_2 \subseteq u^3.$$

We revisit our chart.

2,3?	**3**	**3**	1,3?
[1,3]	**2**	**2**	[2]
[2]	**1**	**1**	[3]
	C_1	C_2	

From this we see that if v^3 does not meet $u^3 \cap C_1$ then $v^2 \cap C_1$ and $v^2 \cap C_2$ are both u-definable, contradicting Lemma 19.3.23, part (1).

Thus v^3 meets $u^3 \cap C_1$ and does not meet $u^3 \cap C_2$. But then the partition of C_2 associated to u is v-definable, and we have a contradiction to Lemma 19.3.23, part (2). $\qquad\square$

19.3.4. Proof that $n = \infty$. It will be considerably easier to prove that n is infinite (Proposition 19.3.26 below).

LEMMA 19.3.25. *Suppose that Γ is a locally degenerate imprimitive homogeneous 3-multi-graph with $\Gamma_1 \cong K_\infty^3[K_n^2]$, and that Γ realizes all triangle types which are not monochromatic. Suppose that $n < \infty$. Then the following hold.*

1. *For $a \in \Gamma_1$, we have $a^1 \cap \Gamma_2 \cong K_{n-1}^2$.*
2. *$n \geq 3$.*
3. *For $u \in \Gamma_2$, the set $u^1 \cap \Gamma_1$ is contained in a single $\overset{2}{\sim}$-class $C(u)$ of Γ_1.*
4. *For $u, v \in \Gamma_2$ distinct, we have $C(u) = C(v)$ iff $u \overset{2}{\rightharpoonup} v$.*

PROOF. We treat each point in turn.

Ad 1. For $a \in \Gamma_1$, the $\overset{2}{\sim}$-class of v_* in a^1 is $\{v_*\} \cup (a^1 \cap \Gamma_2)$, so point (1) is immediate.

Ad 2. Suppose $n = 2$. Then by (1) we may define a function $f : \Gamma_1 \to \Gamma_2$ by $a \overset{1}{\rightharpoonup} f(a)$. As Γ_1 realizes only two non-trivial 2-types it follows that f is a bijection and Γ_2 omits the type 2, that is, there are no monochromatic triangles of type 2. In this case we must have $\Gamma_2 \cong K_\infty^3[K_2^1]$ with f carrying $\overset{2}{\sim}$-classes in Γ_1 to $\overset{1}{\sim}$-classes in Γ_2. For u in Γ_1 or Γ_2 let u' be the other vertex in its $\overset{2}{\sim}$-class or $\overset{1}{\sim}$-class, respectively, and let u^* be its image or preimage under f, respectively.

As there is no monochromatic triangle of type 2 we find easily that $u \overset{2}{\rightharpoonup} u^{*\prime}$.

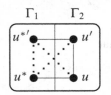

Fix $v \in \Gamma_3$, $u_1 \in v^1 \cap \Gamma_2$, and $u_2 \in v^2 \cap \Gamma_2$. As $u_1^*, u_1', v \in u_1^1$ and $u_2^{*\prime}, v_*, v \in u_2^2$, we find $v \overset{3}{\rightharpoonup} u_1^*, u_1', u_2^{*\prime}$. Then comparison of (v, u_1^*) and $(v, u_2^{*\prime})$ gives a contradiction.

Ad 3. For C a $\overset{2}{\sim}$-class of Γ_1, let

$$C^* = \bigcup_{c \in C} (c^1 \cap \Gamma_2).$$

Then $C^* \subseteq \Gamma_2$ is finite and of fixed size.

We consider the collection

$$\{C^* \mid C \in \Gamma_1/\overset{2}{\sim}\}.$$

This family contains an infinite Δ-system, so by homogeneity it is itself a Δ-system. But

$$\bigcap \{ C^* \mid C \in \Gamma_1 / \overset{2}{\sim} \}, = \emptyset$$

so the sets C^* are pairwise disjoint (and also distinct for distinct classes C). This is point (3).

So for $u \in \Gamma_2$ we now define $C(u)$ as the $\overset{2}{\sim}$-class of Γ_1 containing $u^1 \cap \Gamma_1$.

Ad 4. Let $\approx$ be the equivalence relation on Γ_2 defined by

$$C(u) = C(v).$$

We claim that the relation $\approx$ is $\overset{2}{\sim}$.

For $a \in \Gamma_1$ and $u \in a^1 \cap \Gamma_2$ we have $C(u) = a / \overset{2}{\sim}$. As $a^1 \cap \Gamma_2$ is a non-trivial 2-clique, it follows that the relation $\overset{2}{\sim}$ on Γ_2 is contained in $\approx$.

Now suppose that $\approx$ contains another non-trivial 2-type p. Then $\approx$ is a congruence with a unique type transversal to the equivalence classes, and with infinite quotient.

We conclude that the type $\overset{1}{\text{---}}$ must be contained in $\approx$. From this we will derive a contradiction.

Set $k = |u^1 \cap \Gamma_1|$ for $u \in \Gamma_2$. Note that $k \leq n$ by (3). In fact, taking $u, v \in \Gamma_2$ with $u \overset{1}{\text{---}} v$, since $u^1 \cap v^1 = \emptyset$ we find that

$$k \leq \lfloor n/2 \rfloor \leq n - 2.$$

We divide into two cases.

Case 1. $k > 1$.

We take a $\overset{2}{\sim}$-class C in Γ_1 and consider triples of k-subsets A_1, A_2, A_3 of C with

$$A_i \cap A_j \neq \emptyset$$

for i, j distinct. Then for elements $u_1, u_2, u_3 \in \Gamma_2$ with $u_i^1 \cap \Gamma_1 = A_i$, we have $u_i \overset{2}{\text{---}} u_j$. Thus the type of the triple (u_1, u_2, u_3) over the basepoint is uniquely determined, while the cardinality $|A_1 \cup A_2 \cup A_3|$ may be $k + 1$ or $k + 2$. This is a contradiction.

Case 2. $k = 1$.

Then there is a function $f : \Gamma_2 \to \Gamma_1$ defined by

$$u \overset{1}{\text{---}} f(u).$$

Furthermore, $f(u) = f(v)$ when $u \overset{2}{\text{---}} v$. Thus $u \overset{1}{\text{---}} v$ when $C(u) = C(v)$ and $f(u) \neq f(v)$.

But $n \geq 3$. Take three elements u_1, u_2, u_3 in the same $\overset{2}{\sim}$-class of Γ_1, and take $v_i \in \Gamma_2$ with $f(v_i) = u_i$ to get a monochromatic triangle of type 1, and a contradiction. This completes the proof. $\square$

Proposition 19.3.26. *Suppose that Γ is a locally degenerate imprimitive homogeneous 3-multi-graph with $\Gamma_1 \cong K_m^3[K_n^2]$, and that Γ realizes all triangle types which are not monochromatic. Then $m = n = \infty$.*

Proof. Proposition 19.3.24 shows

$$m = \infty.$$

Suppose toward a contradiction that n is finite.

We use the notation and information from Lemma 19.3.25. Thus for $u \in \Gamma_2$ we let $C(u)$ be the $\overset{2}{\sim}$-class of Γ_1 containing $u^1 \cap \Gamma_1$, and we set

$$k = |u^1 \cap \Gamma_1|.$$

Since $u^1 \cap \Gamma_1 \subseteq C(u)$ we have

$$k \leq n.$$

Recall that for $u, v \in \Gamma_2$ distinct, we have $C(u) = C(v)$ iff $u \overset{2}{\sim} v$.

Claim 1. $k = n$.

Fix a $\overset{2}{\sim}$-class C in Γ_1. Let $C^* = \{u \in \Gamma_2 \mid C(u) = C\}$. Then C^* is a 2-clique.

Consider the function

$$A : C^* \to \mathcal{P}(C)$$

defined by

$$A(u) = u^1 \cap C.$$

If A is not 1-1 on C^* then as C^* is a $\overset{2}{\sim}$-clique, A must be constant on C^*, and thus $A(u) = C$ for $u \in C^*$, and $k = n$.

So suppose the function A is 1-1 on C^*. In particular $k < n$, and we aim at a contradiction.

For $a \in C$ and $a^1 \cap C^*$ distinct, the set $A(u)$ meets $A(v)$. As $A(u) \neq A(v)$ we conclude

$$k > 1.$$

If

$$2 \leq k < n - 1,$$

then for any three distinct k-subsets A_1, A_2, A_3 of C and any u_1, u_2, u_3 in C^* with $A(u_i) = A_i$, as C^* is a $\overset{2}{\sim}$-clique the type of (u_1, u_2, u_3) is uniquely determined. But under our hypotheses on k, the cardinality of $|A_1 \cup A_2 \cup A_3|$ is not uniquely determined. This is a contradiction. So there remains only one case to be considered:

$$k = n - 1.$$

In this case, consider the function $f : \Gamma_2 \to \Gamma_1$ defined by

$$\{f(u)\} = C(u) \setminus A(u).$$

Since A is 1-1 on each set C^*, the function f is a bijection. But Γ_1 realizes two non-trivial 2-types while Γ_2 realizes three.

CLAIM 2. Γ_2 is imprimitive with $\overset{2}{\sim}$-cliques of order $n-1$ as equivalence classes, and with any pair of distinct classes realizing both types $\overset{1}{-}$ and $\overset{3}{-}$.

Fix a $\overset{2}{\sim}$-class C of Γ_1 and consider the associated $\overset{2}{\sim}$-class C^* of Γ_2. We know $k = n$, that is

$$C^* \overset{1}{-} C.$$

Hence $|C^*| = n - 1$ (Lemma 19.3.25, point (1)).

Since we have a bijection $\Gamma_1/\overset{2}{\sim} \leftrightarrow \Gamma_2/\overset{2}{\sim}$ and $\Gamma_1/\overset{2}{\sim}$ realizes a unique non-trivial 2-type, all non-trivial 2-types other than $\overset{2}{-}$ must be realized between any pair of $\overset{2}{\sim}$-classes in Γ_2.

This proves the claim.

CLAIM 3. Let $v \in \Gamma_2$. Then for any non-trivial 2-type p, $v^p \cap \Gamma_1$ is a union of $\overset{2}{\sim}$-classes of Γ_1.

By Claim 1, $v^1 \cap \Gamma_1$ is a $\overset{2}{\sim}$-class of Γ_1.

Next we consider the type 3. Take $u \in \Gamma_2$ with $u \overset{1}{-} v$. In view of the structure of u^1 we have $v \overset{3}{-} C(u)$. Thus $v^3 \cap \Gamma_1$ contains a $\overset{2}{\sim}$-class of Γ_1, and it follows by homogeneity that $v^3 \cap \Gamma_1$ is a union of $\overset{2}{\sim}$-classes of Γ_1.

Since $v^p \cap \Gamma_1$ is a union of $\overset{2}{\sim}$-classes of Γ_1 for $p = 1$ or 3, the same follows also for $p = 2$.

This proves the claim.

Now to conclude, take $v \in \Gamma_2$. We know that v^2 contains $\overset{2}{\sim}$-classes of order n in Γ_1, and Γ_2 has $\overset{2}{\sim}$-classes of order $n-1$, contradicting homogeneity. $\square$

19.3.5. The case $m = n = \infty$: structure of $u^p \cap \Gamma_1$ for $u \in \Gamma_2$. Now we begin the analysis of locally degenerate imprimitive homogeneous 3-multi-graphs in the most plausible (i.e., the most "generic") case: $m = n = \infty$.

Our target in this subsection is the following.

PROPOSITION 19.3.27. Suppose that Γ is a locally degenerate imprimitive homogeneous 3-multi-graph with

$$\Gamma_1 \cong K_\infty^3[K_\infty^2],$$

and that Γ realizes all triangle types which are not monochromatic. Then for $u \in \Gamma_2$ and p a non-trivial 2-type we have

$$u^p \cap \Gamma_1 \cong K_\infty^3[K_\infty^2].$$

LEMMA 19.3.28. *Suppose that Γ is a locally degenerate imprimitive homogeneous 3-multi-graph with*

$$\Gamma_1 \cong K_\infty^3[K_\infty^2].$$

Then for $a \in \Gamma_1$ the structure induced on $a^1 \cap \Gamma_2$ is an infinite 2-clique.

PROOF. Take a as the basepoint. Then the original basepoint v_* lies in a^1 and $v_*^2 \cap a^1$ is an infinite 2-clique.   □

LEMMA 19.3.29. *Suppose that Γ is a locally degenerate imprimitive homogeneous 3-multi-graph with*

$$\Gamma_1 \cong K_\infty^3[K_\infty^2],$$

and that Γ realizes all triangle types which are not monochromatic. With $\ell = 2$ or 3, $u \in \Gamma_\ell$, p a non-trivial 2-type, and C a $\overset{2}{\sim}$-class in Γ_1, suppose that u^p meets C.

Then $u^p \cap C$ is infinite.

PROOF.

CLAIM 1. $|u^p \cap \Gamma_1|$ *is infinite.*

Suppose on the contrary that $u^p \cap \Gamma_1$ is finite. Let $C(u)$ be the union of the $\overset{2}{\sim}$-classes of Γ_1 meeting u^p.

If $C(u)$ contains more than one $\overset{2}{\sim}$-class of Γ_1, then the following four relations are all realized by pairs $u, v \in \Gamma_\ell$.

1. $C(u) \cap C(v) = \emptyset$.
2. $C(u)$ meets $C(v)$ and $C(u) \neq C(v)$.
3. $C(u) = C(v)$ and $u^p \cap \Gamma_1$ is disjoint from $v^p \cap \Gamma_1$.
4. $C(u) = C(v)$ and $u^p \cap \Gamma_1$ meets $v^p \cap \Gamma_1$, but the two sets are distinct.

But only three non-trivial 2-types are realized in Γ_2, a contradiction. Thus $C(u)$ consists of a single $\overset{2}{\sim}$-class of Γ_1, containing $u^p \cap \Gamma_1$.

In particular the (Γ_1, Γ_ℓ)-type partition is non-trivial and of the form $(pq|r)$ with $\{p, q, r\} = \{1, 2, 3\}$, where u^p and u^q split the $\overset{2}{\sim}$-class $C(u)$ into a finite and a co-finite piece, and u^r is the union of the other $\overset{2}{\sim}$-classes of Γ_1.

Now if $|u^p \cap \Gamma_1| > 1$ then the following three relations are realized by pairs u, v in Γ_2.

1. $C(u) \neq C(v)$.
2. $C(u) = C(v)$, $u^p \cap \Gamma_1$ disjoint from $v^p \cap \Gamma_1$.
3. $u^p \cap \Gamma_1$ meets $v^p \cap \Gamma_1$ and the two sets are distinct.

So these three relations correspond in some order to the non-trivial 2-types realized in Γ_2.

Therefore the function mapping u to $u^p \cap \Gamma_1$ is injective and identifies Γ_2 with the 2-cliques of some fixed order $k > 1$ in Γ_1. However, as usual, this is not a homogeneous structure for a binary language (cf. Example 19.3.5), and we arrive at a contradiction.

So we find $|u^p \cap \Gamma_1| = 1$. That is, we have a function $f : \Gamma_\ell \to \Gamma_1$ defined by

$$x \xrightarrow{p} f(x).$$

As Γ_ℓ and Γ_1 realize different numbers of 2-types this is not a bijection, and the following relations on Γ_ℓ must correspond to the non-trivial 2-types on Γ_1.

$$R_1 : f(x) = f(y), x \neq y;$$
$$R_2 : C(x) = C(y), f(x) \neq f(y);$$
$$R_3 : C(x) \neq C(y).$$

Which of these relations corresponds to the type 1?

If $u \xrightarrow{1} v$ and $f(u) = f(v)$ then u^p, u^q, u^r meet Γ_1 in the same sets as v^p, v^q, v^r, respectively, giving triangles of types $(p, p, 1)$, $(q, q, 1)$, and $(r, r, 1)$, and, in particular, a triangle of type $(1, 1, 1)$, a contradiction.

If we fix a $\overset{2}{\sim}$-class C_0 of Γ_1 and for each $a \in C_0$ pick some $u_a \in f^{-1}(a)$, then the u_a form an infinite clique with respect to the edge relation R_2, so the relation R_2 cannot correspond to the type 1.

On the other hand, if we pick representatives for the sets $f^{-1}(C')$ with C' now varying over the $\overset{2}{\sim}$-classes of Γ_1, then we get an infinite clique with respect to the edge relation R_3, so the relation R_3 cannot correspond to the type 1.

Thus we reach a contradiction, and finally prove our claim.

Now suppose toward a contradiction that $k = |u^p \cap C|$ satisfies

$$1 \leq k < \infty.$$

Then for any $\overset{2}{\sim}$-class C' of Γ_1. $|u^p \cap C'|$ is 0 or k.

In particular, by Claim 1 the set u^p meets infinitely many $\overset{2}{\sim}$-classes of Γ_1.

CLAIM 2. *For $u \in \Gamma_\ell$, u^1 meets every $\overset{2}{\sim}$-class of Γ_1.*

Otherwise the (Γ_1, Γ_ℓ)-type partition has the form $(p, q|r)$. That is, u^q is $C(u) \setminus u^p$ and u^r is $\Gamma_1 \setminus C(u)$.

In particular for $u, v \in \Gamma_\ell$ and C a $\overset{2}{\sim}$-class of Γ_1 meeting u^p and v^p we have a point $c \in C$ with $\text{tp}(u, c) = \text{tp}(v, c) = q$. Take u, v so that u^p and v^p have a common point in one $\overset{2}{\sim}$-class C_1 and meet another $\overset{2}{\sim}$-class C_2 in non-empty but disjoint sets. Take points $c_1 \in C_1$, $c_2 \in C_2$ with $\text{tp}(c_i, u) = \text{tp}(c_i, v) = q$. Then the isomorphism $(c_1, u, v) \cong (c_2, u, v)$, with basepoint fixed, contradicts homogeneity.

This proves the claim.

CLAIM 3. *For any $u, v \in \Gamma_\ell$ and any $\overset{2}{\sim}$-class C in Γ_1 the partition of C into three pieces corresponding to u or to v is the same modulo finite sets, and up to the order of the pieces.*

The partition in question consists of $u^p \cap C$, $u^q \cap C$, and $u^r \cap C$ with p, q, r the three non-trivial 2-types. Here $u^p \cap C$ is finite, and if one of the other pieces is also finite then the claim is vacuous. So assume that $u^q \cap C$ and $u^r \cap C$ are infinite.

Take two $\overset{2}{\sim}$-classes C_1, C_2 in Γ_1 and as in the proof of the previous claim take $v \in \Gamma_\ell$ so that $v^p \cap C_1 = u^p \cap C_1$ and $v^p \cap C_2 \neq u^p \cap C_2$. If there are points $c_1 \in C_1$ and $c_2 \in C_2$ so that the types of c_1 and c_2 over u, v agree then we have a contradiction.

In particular if $u^q \cap C_1$ meets both v^q and v^r we arrive at a contradiction, and similarly if $u^r \cap C_1$ meets both v^q and v^r. So after removing $(u^p \cup v^p) \cap C_1$ the partitions agree up to the labels on the sets. This proves the claim.

CLAIM 4. *There is a second non-trivial 2-type, which we will call q, such that u^q meets each $\overset{2}{\sim}$-class of Γ_1 in a finite set.*

This follows from the previous claim. If $u^q \cap C_1$ and $u^r \cap C_1$ are both infinite then we can find $v \in \Gamma_\ell$ so that v^q and v^r intersect $u^q \cap C_1$ in arbitrarily large sets, and hence we may find v so that these intersections are infinite, contradicting the previous claim.

This proves the claim.

Now we reach a contradiction similar to those just seen. We have a 2-type r for which $u^r \cap C$ is cofinite in every $\overset{2}{\sim}$-class of Γ_1, for any $u \in \Gamma_\ell$. So for any pair of vertices $u, v \in \Gamma_\ell$ and any $\overset{2}{\sim}$-class C of Γ_1, the type $\mathrm{tp}(c, u) = \mathrm{tp}(c, v) = r$ is realized in C. Then taking two such classes C_1, C_2 and arranging $u^p \cap C_1 = v^p \cap C_1$ while $u^p \cap C_2 \neq v^p \cap C_2$ we contradict homogeneity. □

LEMMA 19.3.30. *Suppose that Γ is a locally degenerate imprimitive homogeneous 3-multi-graph with*

$$\Gamma_1 \cong K_\infty^3[K_\infty^2],$$

and that Γ realizes all triangle types which are not monochromatic. With $\ell = 2$ or 3 and p a non-trivial 2-type, suppose that u^p meets a finite number of $\overset{2}{\sim}$-classes of Γ_1 for $u \in \Gamma_\ell$. Let

$$C(u) = \bigcup[(u^p \cap \Gamma_1)/\overset{2}{\sim}].$$

Then we have the following.

1. *There are $u, v \in \Gamma_\ell$ distinct such that $C(u) = C(v)$.*
2. *$C(u)$ is a single $\overset{2}{\sim}$-class of Γ_1.*

PROOF. $C(u)$ is the union of the set of $\overset{2}{\sim}$-classes of Γ_1 which meet u^p.

Let k be the number of $\overset{2}{\sim}$-classes of Γ_1 which meet u^p and define a function

$$\gamma : \Gamma_\ell \to \binom{\Gamma_1/\overset{2}{\sim}}{k}$$

by

$$\gamma(u) = C(u)/\overset{2}{\sim},$$

that is, $\gamma(u)$ is the set of $\overset{2}{\sim}$-classes in Γ_1 which meet u^p.

Ad 1. Suppose the contrary. Then the function γ is a bijection.

Since Γ_ℓ is binary homogeneous with three non-trivial 2-types, while $\binom{\Gamma_1/\overset{2}{\sim}}{k}$ is either not binary homogeneous (when $k \geq 2$) or has only one non-trivial 2-type (when $k = 1$), this is a contradiction. Cf. Example 19.3.5.

Ad 2. Suppose $k \geq 2$. Consider the three relations on Γ_ℓ defined as follows.

$$R_1 : \gamma(x) = \gamma(y), x \neq y;$$
$$R_2 : \gamma(x) \cap \gamma(y) \neq \emptyset, \gamma(x) \neq \gamma(y);$$
$$R_3 : \gamma(x) \cap \gamma(y) = \emptyset.$$

These must correspond to the three non-trivial 2-types in Γ_ℓ.

So the equivalence relation $\approx$ on Γ_ℓ defined by $\gamma(x) = \gamma(y)$ must be a congruence and now the induced map

$$\bar{\gamma} : \Gamma_2/\approx \to \binom{\Gamma_1/\overset{2}{\sim}}{k}$$

takes a binary homogeneous structure with two 2-types to $\binom{\Gamma_1/\overset{2}{\sim}}{k}$, giving much the same contradiction as previously. $\square$

LEMMA 19.3.31. *Suppose that Γ is a locally degenerate imprimitive homogeneous 3-multi-graph with*

$$\Gamma_1 \cong K_\infty^3[K_\infty^2],$$

and that Γ realizes all triangle types which are not monochromatic. With $\ell = 2$ or 3, and p a non-trivial 2-type, suppose that $u \in \Gamma_\ell$ and $u^p \cap \Gamma_1$ is a proper subset of a $\overset{2}{\sim}$-class $C(u)$ of Γ_1. Then

1. $u^1 \cap \Gamma_1 \subseteq C(u)$.

Now suppose that the index ℓ is 2. Then we also have the following.

2. *The equivalence relation $\approx$ defined on Γ_2 by $C(x) = C(y)$ is either $\overset{2}{\sim}$ or $\overset{1,2}{\sim}$.*

3. *If $\approx$ is $\overset{2}{\sim}$ then Γ_2 is the generic K_3^1-restricted imprimitive homogeneous 3-multi-graph.*

4. *If $\approx$ is $\overset{1,2}{\sim}$ then Γ_2 is the composition $K^3_\infty[H_3^{1,2}]$ where the superscripts indicate the language used and in particular $H_3^{1,2}$ is the generic triangle free (Henson) graph in a language in which $\overset{1}{\rule{1em}{0.4pt}}$ is the edge relation and $\overset{2}{\rule{1em}{0.4pt}}$ is the non-edge relation.*

PROOF. We treat each point in turn.

Ad 1. We claim u^1 meets $C(u)$. Otherwise, u^2 and u^3 split $C(u)$ and $u^1 \cap \Gamma_1 = \Gamma_1 \setminus C(u)$.

But then if we take $u, v \in \Gamma_\ell$ with $u \overset{1}{\rule{1em}{0.4pt}} v$ and $a \in \Gamma_1 \setminus [C(u) \cup C(v)]$, then the triangle (a, u, v) has type $(1, 1, 1)$, a contradiction.

This proves point (1).

From this, it follows that u^1 meets $C(u)$ in a proper subset (either $u^p \cap \Gamma_1$ or the complement in $C(u)$), and that for some type $q \neq 1$ or p the set $u^q \cap \Gamma_1$ is $\Gamma_1 \setminus C(u)$. In particular we may take the type p to be the type 1 here.

Now we suppose

$$\ell = 2.$$

Ad 2. Take a $\overset{2}{\sim}$-class C of Γ_1 and fix $a \in C$. Then $a^1 \cap \Gamma_2$ is a 2-clique (Lemma 19.3.28) and for $u \in a^1 \cap \Gamma_2$ we have $C(u) = C$. Hence $\overset{2}{\rule{1em}{0.4pt}}$ is contained in $\approx$.

As $\approx$ is a non-trivial equivalence relation, it is $\overset{2}{\sim}$, $\overset{1,2}{\sim}$, or $\overset{2,3}{\sim}$.

If $\approx$ is $\overset{2,3}{\sim}$, then $\overset{1}{\rule{1em}{0.4pt}}$ is transversal to $\approx$ and there is an infinite 1-clique, for a contradiction. Thus $\approx$ must be $\overset{2}{\sim}$ or $\overset{1,2}{\sim}$.

Thus the second point is proved.

Now we consider the bijection

$$\bar{\gamma} : \Gamma_2/\approx \; \cong \; \Gamma_1/\overset{2}{\sim}$$

induced by $\gamma(u) = C(u)$. This is definable from the basepoint, and at this stage we work systematically over the basepoint, which is treated as a constant in the language.

When $\approx$ is not a congruence we must be a little careful about the nature of the structure $\Gamma_2/\approx$; but what matters is the permutation group induced on this structure by the stabilizer of the basepoint in Aut(Γ). We continue to refer to its orbits on finite sequences in $\Gamma_2/\approx$ as *types*; these correspond via $\bar{\gamma}$ to types in the usual sense in $\Gamma_1/\overset{2}{\sim}$, which is an infinite set with no additional structure.

Ad 3. We suppose $\approx$ is $\overset{2}{\sim}$.

We know that Γ_2 is an imprimitive homogeneous 3-multi-graph with equivalence relation $\overset{2}{\sim}$ and infinite 2-cliques, omitting K^1_3, and with the quotient $\Gamma_2/\overset{2}{\sim}$ infinite, and realizing only one non-trivial 2-type (the induced permutation group is doubly transitive).

Going through the classification of the imprimitive homogeneous 3-multi-graphs we find the following possibilities; we will eliminate all but the generic 3-multi-graph with $\overset{2}{\sim}$ an equivalence relation and K_3^1 forbidden.

1(a) Composite: The types 1, 3 must occur between each pair of equivalence classes. So Γ_2 is not composite.

1(b) Product: If Γ is a product with the stated properties, then it contains an infinite clique for each non-trivial 2-type, a contradiction.

1(c) Double Cover: Γ_2 contains infinite 2-cliques.

1(d) Clique-restricted: This is the case envisaged, and we know there are infinitely many equivalence classes, so there is just one possibility of this type.

1(e, f) Semi-generic or generic imprimitive: This will contain an infinite 1-clique.

Ad 4. Now we suppose that the relation $\approx$ is $\overset{1,2}{\sim}$, and in particular this relation is a congruence. So $\Gamma_2 \cong K_\infty^3[\Delta]$ where Δ is the structure induced on a single $\overset{1,2}{\sim}$-class.

If Δ is primitive, then it is the generic triangle-free graph, with $\overset{1}{-}$ taken as the edge relation, as claimed.

If Δ is imprimitive then it is one of the following.

$$K_\infty^2[K_2^1], \qquad\qquad K_2^1[K_\infty^2].$$

We must eliminate these possibilities.

We take $u \in \Gamma_2$ and consider the structure of u^2.

Case 1. $\Delta \cong K_\infty^2[K_2^1]$.

In u^2 we have

$$v_* \overset{1}{-} u^2 \cap \Gamma_1.$$

In view of the assumed structure of Γ_2 this says that $u^2 \cap \Gamma_1$ is a single point, contradicting Lemma 19.3.29.

Case 2. $\Delta \cong K_2^1[K_\infty^2]$.

In u^2, the $\overset{1,2}{\sim}$-class of v_* is $u^2 \cap (\Gamma_1 \cup \Gamma_2)$. So this is the union of two 2-cliques. We conclude that

$$u^2 \cap \Gamma_1 \overset{1}{-} u^2 \cap \Gamma_2.$$

In particular $u^2 \cap \Gamma_1 \subseteq C(u)$.

Take $v \in u^1 \cap \Gamma_2$. Then $C(v) = C(u)$ and v^1, v^2 also splits $C(u)$. As $u^2 \cap \Gamma_2 \overset{1}{-} u^2 \cap \Gamma_1, v$, it follows that v^1 does not meet $u^2 \cap \Gamma_1$, and hence

$$v \overset{2}{-} u^2 \cap \Gamma_1.$$

So $u^2 \cap \Gamma_1 \subseteq v^2 \cap \Gamma_1$. By symmetry

$$u^2 \cap \Gamma_1 = v^2 \cap \Gamma_1.$$

As the $\approx$-class C_u of u in Γ_2 is complete bipartite, $x^2 \cap \Gamma_1$ is independent of x for $x \in C_u$. But then $\bigcup_{x \in \Gamma_2} x^2 \cap \Gamma_1 \neq \Gamma_1$, a contradiction. $\qquad\square$

LEMMA 19.3.32. *Suppose that Γ is a locally degenerate imprimitive homogeneous 3-multi-graph with $\Gamma_1 \cong K_\infty^3[K_\infty^2]$, and that Γ realizes all triangle types which are not monochromatic. Suppose that $u^1 \cap \Gamma_1$ is a proper subset of a $\overset{2}{\sim}$-class $C(u)$ of Γ_1 for $u \in \Gamma_2$. Then on Γ_2 we have $C(x) = C(y)$ iff $x \overset{1,2}{\sim} y$.*

PROOF. By Lemma 19.3.31 the alternative is that $C(x) = C(y)$ iff $x \overset{2}{\sim} y$, in which case Γ_2 is the generic imprimitive homogeneous 3-multi-graph with monochromatic triangles of type 1 forbidden, where the equivalence relation in question is $\overset{2}{\sim}$.

CLAIM 1. *For u in Γ_2, $u^1 \cap C(u)$ and $u^2 \cap C(u)$ partition $C(u)$.*

We consider $u^1 \cong K_\infty^3[K_\infty^2]$.

We know $u^1 \cap \Gamma_1$ is an infinite 2-clique and $u^1 \cap \Gamma_2$ is $K_\infty^3[K_\infty^2]$. If $u^1 \cap \Gamma_1$ is in the same $\overset{2}{\sim}$-class of u^1 as one of the classes meeting $u^1 \cap \Gamma_2$, then the corresponding class in $u^1 \cap \Gamma_2$ becomes definable from u in Γ_2, which is not the case. We conclude that

$$u^1 \cap \Gamma_1 \overset{3}{\perp} u^1 \cap \Gamma_2.$$

Fix $v \in u^1 \cap \Gamma_2$. Then v^3 meets $C(u)$. As the relation $\approx$ coincides with $\overset{2}{\sim}$, we also have $C(v) \neq C(u)$

Therefore $C(u) \subseteq v^3$ and the (Γ_1, Γ_2)-type partition is $(12|3)$. This is our claim.

CLAIM 2. *For $u \in \Gamma_2$ we have $u^2 \cap \Gamma_1 \overset{1}{\perp} u^2 \cap \Gamma_2$.*

In u^2 we find

$$v_* \overset{1}{\perp} u^2 \cap C(u);$$
$$v_* \overset{2}{\perp} u^2 \cap \Gamma_2.$$

Hence in view of the structure of Γ_2 the only relations between $u^2 \cap C(u)$ and $u^2 \cap \Gamma_2$ are $\overset{1}{\perp}$ and $\overset{3}{\perp}$.

However $C(u) = C(v)$ for $v \in u^2 \cap \Gamma_2$ and hence the only relations holding between these two sets are $\overset{1}{\perp}$ and $\overset{2}{\perp}$.

The claim follows.

Consider $a \in u^2 \cap \Gamma_1$. If $v \in a^2 \cap \Gamma_2$ then $C(v) = C(u)$ and hence $u \overset{2}{\sim} v$. But then by the foregoing if $v \neq u$ we have $v \in a^1$. So $a^2 \cap \Gamma_2 = \{u\}$. Thus we may write $u = f(a)$. Then $f(a)^2 \cap \Gamma_1$ is an infinite a-definable subset of $a/\overset{2}{\sim}$ and hence equals $a/\overset{2}{\sim}$. So $u^2 \cap \Gamma_1 = C(u)$, a contradiction. $\qquad\square$

LEMMA 19.3.33. *Suppose that Γ is a locally degenerate imprimitive homogeneous 3-multi-graph with*

$$\Gamma_1 \cong K^3_\infty[K^2_\infty],$$

and that Γ realizes all triangle types which are not monochromatic. Suppose $u \in \Gamma_2$ and $u^1 \cap \Gamma_1$ meets finitely many $\overset{2}{\sim}$-classes of Γ_1. Then $u^1 \cap \Gamma_1$ is a $\overset{2}{\sim}$-class of Γ_1.

PROOF. By Lemma 19.3.30 $u^1 \cap \Gamma_1$ is contained in a unique $\overset{2}{\sim}$-class $C(u)$ in Γ_1.

Suppose toward a contradiction that $u^1 \cap C(u)$ is a proper subset of $C(u)$. Then by Lemmas 19.3.31 and 19.3.32 we have the following conditions.

(a) For $x, y \in \Gamma_2$, $C(x) = C(y)$ iff $x \overset{1,2}{\sim} y$.
(b) $\Gamma_2 \cong K^3_\infty[H^{1,2}_3]$.

Let D be a $\overset{1,2}{\sim}$-class in Γ_2.

CLAIM 1. *For $u \in D$, $(u^2 \cap D) \overset{3}{-} (u^2 \cap \Gamma_3)$.*

In u^2 we have

$$v_* \overset{2}{-} u^2 \cap \Gamma_2,$$

$$v_* \overset{3}{-} u^2 \cap \Gamma_3,$$

and in view of the structure of u^2 the claim follows.

Now Claim 1 implies that for $v \in \Gamma_3$, $v^2 \cap D$ is a 1-clique and hence has order at most 2.

CLAIM 2. *For $v \in \Gamma_3$ and any non-trivial 2-type p, v^p meets every $\overset{1,2}{\sim}$-class of Γ_2.*

It suffices to find one such $v \in \Gamma_3$. Take $v \in u^2 \cap \Gamma_3$. Then v^2 and v^3 meet D, by Claim 1. If v^1 meets D as well then the claim follows by homogeneity.

Otherwise, v^2, v^3 split D and v^1 contains a $\overset{1,2}{\sim}$-class, which gives a monochromatic triangle of type 1, and a contradiction.

CLAIM 3. *For $v \in \Gamma_3$, v^2 meets each $\overset{1,2}{\sim}$-class of Γ_2 in a single point.*

Otherwise, $v^2 \cap D$ consists of a pair of points u_1, u_2 with $u_1 \overset{1}{-} u_2$. Then every other vertex u of D lies in $u_1^2 \cup u_2^2$, so by Claim 1 we have $v \overset{3}{-} D \setminus \{u_1, u_2\}$. But also v^1 meets D and so we have a contradiction.

CLAIM 4. *For $v_1, v_2 \in \Gamma_3$ and any two $\overset{1,2}{\sim}$-classes D_1, D_2 in Γ_2, the pair (v_1, v_2) realizes the same type over D_1 or D_2; that is, there is an automorphism fixing the basepoint and taking (v_1, v_2, D_1) to (v_1, v_2, D_2).*

It suffices to find points $d_1 \in D_1$, $d_2 \in D_2$ realizing the same types over v_1, v_2. We take $d_1 \overset{2}{=} v_1^2 \cap D_1, v_2^2 \cap D_1$ and similarly for d_2. By Claim 1 we have $v_1, v_2 \overset{3}{=} d_1, d_2$ and so (v_1, v_2, d_1) and (v_1, v_2, d_2) have the same type. This proves the claim.

Now we arrive at a contradiction. We take $\overset{1,2}{\sim}$-classes D_1, D_2 in Γ_2 and vertices $d_1, d_1' \in D_1$, $d_2, d_2' \in D_2$ so that $d_i \overset{i}{\perp} d_i'$ for $i = 1, 2$. We take vertices $v, v' \in \Gamma_3$ so that $v \overset{2}{=} d_1, d_2$ and $v' \overset{2}{=} d_1', d_2'$.

Then v, v' realize different types over D_1 and D_2, namely

$$v \cap D_1 \overset{1}{\perp} v' \cap D_1;$$
$$v \cap D_2 \overset{2}{=} v' \cap D_2.$$

This contradicts the previous claim.

As we have arrived at a contradiction, we conclude that $u^1 \cap \Gamma_1 = C(u)$. $\square$

Lemma 19.3.34. *Suppose that Γ is a locally degenerate imprimitive homogeneous 3-multi-graph with*

$$\Gamma_1 \cong K_\infty^3[K_\infty^2]$$

and that Γ realizes all triangle types which are not monochromatic.

Suppose that for $u \in \Gamma_2$ the set $C(u) = u^1 \cap \Gamma_1$ is a $\overset{2}{\sim}$-class. Then the following hold.

1. *On Γ_2 we have $C(u) = C(v) \iff u \overset{2}{\sim} v$.*
2. *For $u \in \Gamma_2$ and C a $\overset{2}{\sim}$-class in Γ_1, if u^2 meets C then u^2 contains C.*
3. *Γ_2 is a generic K_3^1-restricted imprimitive homogeneous 3-multi-graph.*
4. *If $u, v \in \Gamma_2$ and $u \overset{2}{=} v$ then the sets $u^2 \cap \Gamma_1$ and $v^2 \cap \Gamma_1$ are disjoint.*
5. *For D a $\overset{2}{\sim}$-class in Γ_2, C the corresponding $\overset{2}{\sim}$-class in Γ_1, and a in $\Gamma_1 \setminus C$, $a^2 \cap \Gamma_2$ is a transversal to $\overset{2}{\sim}$ in $\Gamma_2 \setminus D$.*

Proof. We treat each point in turn.

Ad 1. For $a \in \Gamma_1$ we know $a^2 \cap \Gamma_2$ is a non-trivial 2-clique by Lemma 19.3.28. So if $C(u) = C(v)$ then $u \overset{2}{\sim} v$, and the converse follows by homogeneity.

In particular $\overset{2}{\sim}$ is an equivalence relation on Γ_2.

Ad 2. Suppose $a \in C$ and $a \overset{2}{=} u$. Taking a as basepoint, in a^2 we either have $u \overset{2}{=} C \setminus \{a\}$, in which case the claim holds, or $u^2 \cap C \setminus \{a\}$ is empty, and $u^2 \cap C$ is finite, which contradicts Lemma 19.3.29.

Ad 3. This follows as in the proof of Lemma 19.3.31.

Ad 4. Take $a \in u^2 \cap \Gamma_1$. The structure induced on u^2 is a generic K_3^1-restricted imprimitive homogeneous 3-multi-graph with equivalence relation $\overset{2}{\sim}$. In u^2 we have

$$v_* \overset{1}{\perp} a; \qquad\qquad v_* \overset{2}{=} v.$$

So the type of av is 1 or 3, and the claim follows.

Ad 5. It follows from the point (4) that a^3 meets D. Hence there is a unique type of pair (a, D) of this kind. Since we have such pairs a, D for which a^2 meets D, we find that a^2 meets each $\overset{2}{\sim}$-class of $\Gamma_2 \setminus D$.

By point (4), a^2 meets each $\overset{2}{\sim}$-class of Γ_2 in at most one vertex. The claim follows. □

PROOF OF PROPOSITION 19.3.27. Suppose that $u \in \Gamma_2$ and $u^p \cap \Gamma_1$ is not isomorphic to $K_\infty^3[K_\infty^2]$.

Then by Lemmas 19.3.29 and 19.3.30, $u^p \cap \Gamma_1$ meets a unique $\overset{2}{\sim}$-class. By Lemma 19.3.31 we may suppose the type p is type 1. By Lemma 19.3.33, $u^1 \cap \Gamma_1$ is a $\overset{2}{\sim}$-class $C(u)$.

We then have the situation described in Lemma 19.3.34. In particular, for D a $\overset{2}{\sim}$-class in Γ_2 there is a unique $\overset{2}{\sim}$-class $C(D)$ in Γ_1 with

$$C(D) = D^1 \cap \Gamma_1.$$

For $C \neq C(D)$ any other $\overset{2}{\sim}$-class in Γ_1 we have

$$C \overset{2}{\_} u \text{ for a unique } u = u(C, D) \in D,$$
$$C \overset{3}{\_} D \setminus \{u(C, D)\}.$$

Now we have distinct pairs C_1, C_2 with $u(C_1, D) = u(C_2, D)$, and also pairs C_1', C_2' with $u(C_1', D) \neq u(C_2', D)$.

We fix representatives c_1, c_2, c_1', c_2'. Then we may find $u \in D$ with $u \overset{3}{\_} c_1, c_2, c_1', c_2'$. It follows that (u, c_1, c_2) is isomorphic to (u, c_1', c_2'). This violates homogeneity. □

19.3.6. The (Γ_1, Γ_2)-type partition. We aim at the following.

PROPOSITION 19.3.35. *Suppose that Γ is a locally degenerate imprimitive homogeneous 3-multi-graph with*

$$\Gamma_1 \cong K_\infty^3[K_\infty^2]$$

and that Γ is not of known type. Then for every $u \in \Gamma_2$ and every $\overset{2}{\sim}$-class C of Γ_1, all three non-trivial 2-types are realized in C over u. In other words, the (Γ_1, Γ_2)-type partition has one class.

LEMMA 19.3.36. *Suppose that Γ is a locally degenerate imprimitive homogeneous 3-multi-graph with $\Gamma_1 \cong K_\infty^3[K_\infty^2]$, and that Γ is not of known type. Then $\overset{2}{\sim}$ is the only non-trivial, proper equivalence relation on Γ_2.*

In particular, Γ_2 is 1-connected (that is, connected as a graph with edge relation $\overset{1}{\_}$).

PROOF. By Proposition 19.3.27 $u^2 \cap \Gamma_1 \cong K_\infty^3[K_\infty^2]$.

Take $u \in \Gamma^2$. In u^2 we have

$$v_* \xrightarrow{\ 1\ } u^2 \cap \Gamma_1.$$

This gives triangles of types $(1, 1, 2)$ and $(1, 1, 3)$ in u^2, hence also in Γ_2. In addition, there are triangles of type $(3, 3, 2)$ in $u^2 \cap \Gamma_1$, hence in Γ_2.

So any non-trivial and proper equivalence relation on Γ_2 is contained in $\overset{2,3}{\sim}$ and contains $\overset{2}{\sim}$. In particular Γ_2 is 1-connected.

Suppose now that $\overset{2,3}{\sim}$ is an equivalence relation on Γ_2. Then $\Gamma_2/\overset{2,3}{\sim} \cong K_2^1$. Let A, B be the two $\overset{2,3}{\sim}$-classes of Γ_2. Then

$$A \xrightarrow{\ 1\ } B.$$

Let $A^* = \bigcup_{u \in A} u^1 \cap \Gamma_1$ and $B^* = \bigcup_{u \in B} u^1 \cap \Gamma_1$. Then (A^*, B^*) is a v_*-definable partition of Γ_1. This is a contradiction.

So the only non-trivial, proper equivalence relation on Γ_2 is $\overset{2}{\sim}$. □

We now consider the (Γ_1, Γ_2)-type partition associated with Γ. Two non-trivial 2-types p_1, p_2 are in the same class for this partition if they occur in a configuration $a_1 \xrightarrow{\ p_1\ } u$, $a_2 \xrightarrow{\ p_2\ } u$ with $a_1, a_2 \in \Gamma_1$, $a_1 \overset{2}{\sim} a_2$, and $u \in \Gamma_2$.

LEMMA 19.3.37. *Suppose that Γ is a locally degenerate imprimitive homogeneous 3-multi-graph with*

$$\Gamma_1 \cong K_\infty^3[K_\infty^2]$$

and that Γ is not of known type. Then in the (Γ_1, Γ_2)-type partition, $\xrightarrow{\ 1\ }$ is associated with at least one other 2-type.

PROOF. We suppose the contrary and derive a contradiction from an amalgamation argument.

CLAIM 1. Γ *contains a configuration $av_1v_2v_3$ with*

$$a \xrightarrow{\ 1\ } v_1, v_2; \qquad\qquad\qquad a \xrightarrow{\ 2\ } v_3;$$

$$v_2 \xrightarrow{\ 3\ } v_1, v_3; \qquad\qquad\qquad v_1 \xrightarrow{\ 1\ } v_3.$$

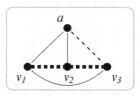

From the point of view of a as basepoint, this configuration consists of $v_3 \in \Gamma_2$, $v_1 \in v_3^1 \cap \Gamma_1$, and $v_2 \in v_3^3 \cap \Gamma_1$, with $v_1 \xrightarrow{\ 3\ } v_2$. This is afforded by Proposition 19.3.27.

Now we perform an amalgamation. We take b with $\mathrm{tp}(b/v_1v_2v_3) = \mathrm{tp}(a/v_3v_2v_1)$ and amalgamate a, b over $v_1v_2v_3$. As $v_2 \xrightarrow{\ 1\ } a, b$ we find $a \xrightarrow{\ i\ } b$

with $i = 2$ or 3. Then taking v_3 as basepoint the configuration shows that the 2-types $\overset{1}{-}$ and $\overset{i}{-}$ fall into the same class. □

LEMMA 19.3.38. *Suppose that Γ is a locally degenerate imprimitive homogeneous 3-multi-graph with*

$$\Gamma_1 \cong K_\infty^3[K_\infty^2],$$

and that Γ is not of known type. Then the (Γ_1, Γ_2)-type partition is not the partition $(12|3)$

PROOF. We suppose the contrary and work toward a contradiction.

CLAIM 1. *There are $a \in \Gamma_1$ and $u, v \in \Gamma_2$ satisfying*

$$a \overset{2}{-} u, \qquad\qquad a \overset{3}{-} v, \qquad\qquad u \overset{1}{-} v.$$

Take $C \in \Gamma_1/\overset{2}{\sim}$ and let

$$C^* = C^{1,2} \cap \Gamma_2.$$

In other words, C^* consists of those $u \in \Gamma_2$ for which the types realized by u over C are $\overset{1}{-}$ and $\overset{2}{-}$.

Now $C^* \neq \Gamma_2$ and Γ_2 is 1-connected by Lemma 19.3.36, so there are $u \in C^*$ and $v \in \Gamma_2 \setminus C^*$ with $u \overset{1}{-} v$. There is $a \in C$ with $a \overset{2}{-} u$ and then a, u, v provide the required configuration.

This proves the claim.

Now we analyze an amalgamation diagram with two factors, $a1234$ and $b1234$, which have the following type structure (shown first as a table, and then as a diagram, the latter given in two parts).

	b	1	2	3	4
a	?	1	3	1	2
1	1		2	3	2
2	1	2		3	1
3	2	3	3		1
4	2	2	1	1	

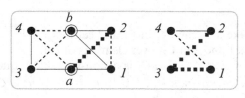

In this diagram, the vertex 1 forces $a \overset{i}{-} b$ with $i = 2$ or 3, and taking vertex 1 or 3 as basepoint in the completed diagram then gives either $2 \overset{1,3}{-} (a, b)$ or $b \overset{2,3}{-} (a, 4)$, so that the type $\overset{3}{-}$ is in the same class as types 1 and 2, for a contradiction. So it suffices to show that the factors $a1234$ and $b1234$ of this amalgamation diagram occur in Γ.

CLAIM 2. *The configuration $a1234$ is the unique amalgam of the configurations $a124$ and $a134$ which does not create a monochromatic triangle of type 1 or put the type $\overset{3}{-}$ in the same class as types 1 and 2.*

Similarly, $b1234$ is the unique amalgam of $b124$ and $b134$ which does not put the type $\overset{3}{-}$ in the same class as types 1 and 2.

For this, consider the resulting configurations 4231 or b123, respectively, where in each case the first vertex listed is to be viewed as the basepoint. The claim follows by inspection.

So it suffices to show that the four configurations a124, a134, b124, and b134 with the structure specified all occur in Γ. Here both of the configurations a124 and b134 are provided by Claim 1.

The configuration a134 with a taken as basepoint consists of the pair 1, 3 in Γ_1 and 4 in Γ_2 with $1 \overset{3}{\longrightarrow} 3$, and is afforded by Proposition 19.3.27. The configuration b124, with b taken as basepoint, consists of a vertex 4 in Γ_2 realizing the types 1 and 2 over a $\overset{2}{\sim}$ class of Γ_1, which we have as the (Γ_1, Γ_2)-type partition is supposed to be $(12|3)$.

Thus the necessary factors are all available, and the amalgamation diagram produces a contradiction. $\square$

Now we can prove the main result of this subsection: each vertex in Γ_2 realizes all non-trivial 2-types over each $\overset{2}{\sim}$-class of Γ_1.

PROOF OF PROPOSITION 19.3.35. We claim that the (Γ_1, Γ_2)-type partition has only one class.

Suppose the contrary. Then from Lemmas 19.3.37 and 19.3.38, the (Γ_1, Γ_2)-type partition is

$$(13|2).$$

So we assume this is the case, and we work toward a contradiction.

CLAIM 1. Γ *contains a configuration with* $a \in \Gamma_1$, $u, v \in \Gamma_2$, *and*

$$a \overset{1}{\longrightarrow} u, \qquad\qquad a \overset{3}{\longrightarrow} v, \qquad\qquad u \overset{2}{\longrightarrow} v.$$

Take a $\overset{2}{\sim}$-class C in Γ_1 and set $C^* = C^{1,3} \cap \Gamma_2$ (the set of realizations of types 1 and 3 in Γ_2, over elements of C).

For $a \in C$, $a^1 \cap \Gamma_2$ is a non-trivial 2-clique (Lemma 19.3.28), so there are elements $u, v \in C^*$ with $u \overset{2}{\longrightarrow} v$. If $u^1 \cap C \neq v^1 \cap C$ then the desired configuration is realized.

Suppose $u^1 \cap C = v^1 \cap C$. Take two elements $a, b \in u^1 \cap C$. Then there are $u_1, v_1 \in \Gamma_2$ for which $a, b \in u_1^1 \cap \Gamma_1$ and $v_1^1 \cap \{a, b\} = \{a\}$. In particular $u_1, v_1 \in a^1 \cap \Gamma_2$ so $u_1 \overset{2}{\longrightarrow} v_1$.

As auv and au_1v_1 are isomorphic there is an automorphism taking Cu_1v_1 to Cuv. But $u^1 \cap C = v^1 \cap C$ while $u_1^1 \cap C \neq v_1^1 \cap C$, so this is a contradiction. This proves the claim.

Now we work toward an amalgamation diagram with two factors a1234 and b1234 with the following type structure, shown as a table and as a diagram.

In this diagram the vertex 1 forces $a \overset{i}{\longrightarrow} b$ with $i = 2$ or 3 and then the configuration 12ab or 34ab with first vertex as basepoint forces $2 \overset{1,2}{\longrightarrow} (a, b)$ or

	b	1	2	3	4
a	?	1	1	2	1
1	1		2	2	3
2	2	2		2	1
3	2	2	2		1
4	2	3	1	1	

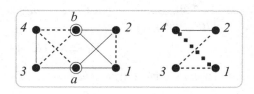

$b \overset{2,3}{\sim} (4, a)$, showing that the type 2 is in the same class of the partition as types 1 and 3.

So it suffices to show that the configurations $a1234$ and $b1234$ occur in Γ.

Now viewing $a123$ as an amalgam of $a13$ and $a23$ forces $1 \overset{2}{\sim} 2$ or $1 = 2$, so $a1234$ is the unique amalgam of $a134$ and $a234$. The configuration $b1234$ is the unique amalgam of $b134$ and 1234 since the configuration $423b$, with 4 as basepoint, forces $b \overset{2}{\sim} 2$. So it suffices to show that the subfactors $a134$, $a234$, $b134$, and 1234 all occur in Γ.

We first consider $a134$ and $a234$. With 3 as basepoint, the configuration $a134$ represents $a, 4$ in Γ_1 and 1 in Γ_2 realizing the types $1, 3$, so this configuration is afforded by the-type partition $(13|2)$. With a as basepoint, the configuration $a234$ consists of 4 in Γ_2 and $2, 3$ in $4^1 \cap \Gamma_1$, so this configuration is realized.

Over the basepoint 3 the configuration 1234 is afforded by Claim 1. This leaves only the configuration $b134$ for consideration.

Over the basepoint 3 this consists of a vertex $4 \in \Gamma_1$ related to a pair of vertices $1, b \in \Gamma_2$ with $1 \overset{1}{\sim} b$ by the types 2 and 3.

Take a $\overset{2}{\sim}$-class C in Γ_1 and let $C^* = C^{1,3} \cap \Gamma_2$. Since Γ_2 is 1-connected there are $u \in C^*$ and $v \in \Gamma_2 \setminus C^*$ with $u \overset{1}{\sim} v$. Take $a \in C$ with $a \overset{3}{\sim} u$. Then $a \overset{2}{\sim} v$ and we have the required configuration. $\square$

19.3.7. Types of $\overset{2}{\sim}$-classes in Γ_1 over pairs in Γ_2. The question that concerns us now is which pairs of types can be realized by a point in a $\overset{2}{\sim}$-class C in Γ_1 over two vertices u, v of Γ_2. As any such pair of types determines the type of the class C over that pair, once the type of u, v is fixed this gives us a partition of the pairs of non-trivial 2-types to consider. For $u, v \in \Gamma_2$ and $a \in \Gamma_1$ with $a \overset{i}{\sim} u, a \overset{j}{\sim} v$ we will refer to the type of a over u, v as (i, j).

We record the point of departure for our analysis as established in previous subsections.

LEMMA 19.3.39. *Suppose that Γ is a locally degenerate imprimitive homogeneous 3-multi-graph with*

$$\Gamma_1 \cong K_\infty^3[K_\infty^2],$$

and that Γ is not of known type. Then for $u \in \Gamma_2$ and C a $\overset{2}{\sim}$-class of Γ_1, the sets

$$u^1 \cap C, u^2 \cap C, u^3 \cap C$$

partition C into three infinite sets.

PROOF. By Proposition 19.3.35 these sets are non-empty and so by Proposition 19.3.27 they are infinite.   □

LEMMA 19.3.40. *Suppose that Γ is a locally degenerate imprimitive homogeneous 3-multi-graph with*

$$\Gamma_1 \cong K_\infty^3[K_\infty^2],$$

and that Γ is not of known type. Then for $u, v \in \Gamma_2$ with $u \overset{2}{\longrightarrow} v$ and C a $\overset{2}{\sim}$-class of Γ_1, every type (i, j) is realized by some $a \in C$ over u, v.

PROOF. Take distinct vertices a_{ij} of C for $i, j \in \{1, 2, 3\}$, and take u_0, v_0 in Γ_2 realizing the types

$$u_0 \overset{i}{\longrightarrow} a_{i,j}, v_0 \overset{j}{\longrightarrow} a_{i,j} \ (i, j \in \{1, 2, 3\}).$$

Since $u^p \cap C$ is infinite for $p \in \{1, 2, 3\}$, such elements u_0, v_0 exist by homogeneity.

As $a_{11} \overset{1}{\longrightarrow} u_0, v_0$ we find $u_0 \overset{2}{\longrightarrow} v_0$ (Lemma 19.3.28). By construction C realizes all types (i, j) over u_0, v_0. For any other pair $u, v \in \Gamma_2$ with $u \overset{2}{\longrightarrow} v$ and any $a \in C$ the type of a over uv is realized by some $b \in C$ over u_0, v_0, and so the same result applies to u, v and C.   □

LEMMA 19.3.41. *Suppose that Γ is a locally degenerate imprimitive homogeneous 3-multi-graph with*

$$\Gamma_1 \cong K_\infty^3[K_\infty^2],$$

and that Γ is not of known type. Then Γ realizes the configuration $a_1 a_2 u_1 u_2$ with $a_1, a_2 \in \Gamma_1, u_1, u_2 \in \Gamma_2, a_i \overset{1}{\longrightarrow} u_i$ for $i = 1, 2$, and all other 2-types type 2.

PROOF. Fix a $\overset{2}{\sim}$-class C in Γ_1 and take $u_1, u_2 \in \Gamma_2$ with $u_1 \overset{2}{\longrightarrow} u_2$. Apply the previous lemma to complete the diagram with $a_1, a_2 \in C$.   □

LEMMA 19.3.42. *Suppose that* Γ *is a locally degenerate imprimitive homogeneous 3-multi-graph with*

$$\Gamma_1 \cong K_\infty^3[K_\infty^2],$$

and that Γ *is not of known type. For C a* $\overset{2}{\sim}$*-class of* Γ_1, $a \in C$, *and* $u, v \in \Gamma_2$ *with* $u \overset{1}{-} v$, *there is* $b \in C$ *satisfying*

$$\mathrm{tp}(b/uv) = \mathrm{tp}(a/vu).$$

PROOF. It suffices to find some pair $a, b \in C$ with $\mathrm{tp}(b/uv) = \mathrm{tp}(a/vu)$, as then we may find an automorphism carrying Cuv to Cvu and fixing the basepoint.

Suppose there is no such pair, and consider the relation $i \to j$ defined on the set $\{1, 2, 3\}$ by the condition

$$u^i \cap v^j \cap C \neq \emptyset.$$

This defines an anti-symmetric, and in particular, irreflexive, relation on the set of non-trivial 2-types. Thus we view this as the arc relation in a directed graph on three vertices with no loops.

By definition, all out-degrees in this directed graph are positive, and thus it is a cyclic tournament. Thus there is a cyclic permutation σ of $\{1, 2, 3\}$ such that

$$v^i \cap C = u^{\sigma(i)} \cap C \ (i = 1, 2, 3).$$

Relabeling u, v as v, u changes the permutation σ to σ^{-1}. Accordingly there is also some $v_0 \in \Gamma_2$ with $u \overset{1}{-} v_0$ and

$$v_0^i \cap C = u^{\sigma^{-1}(i)} \cap C \ (i = 1, 2, 3).$$

For any u', v' in Γ_2 with $u' \overset{1}{-} v'$ and any $\overset{2}{\sim}$-class C' in Γ_1, one of the types $(1, 2)$ or $(1, 3)$ must be realized in C' over u', v' and hence by homogeneity the sets $v'^i \cap C'$ form a cyclic permutation of the sets $u'^i \cap C'$.

If there are u, v_1, v_2 in Γ_2 with

$$u \overset{1}{-} v_1, v_2 \qquad\qquad v_1 \overset{2}{-} v_2$$

then the sets $v_2^i \cap C$ are a cyclic permutation of the sets $v_1^i \cap C$. But this contradicts Lemma 19.3.40. So this configuration does not occur.

On the other hand Γ_2 is 1-connected and hence for $u \in \Gamma_2$ the set $u^1 \cap \Gamma_2$ must be infinite, and thus is an infinite 3-clique. But each element of this clique has a type related to the type of u over C by either σ or σ^{-1} and hence we find $v_1, v_2 \in \Gamma_2$ realizing the same type over C, and with $v_1 \overset{3}{-} v_2$. But taking $a \in C$ with $a \overset{1}{-} v_1, v_2$, we find $v_1 \overset{2}{-} v_2$, a contradiction. $\square$

LEMMA 19.3.43. *Suppose that* Γ *is a locally degenerate imprimitive homogeneous 3-multi-graph with*

$$\Gamma_1 \cong K_\infty^3[K_\infty^2],$$

and that Γ is not of known type. Then Γ realizes the following configuration.

$$a_1, a_2 \in \Gamma_1; \qquad\qquad a_1 \overset{2}{\relbar} a_2;$$

$$u, v \in \Gamma_2; \qquad\qquad u \overset{1}{\relbar} v;$$

$$a_1 \overset{1}{\relbar} u, a_1 \overset{3}{\relbar} v; \qquad\qquad a_2 \overset{2}{\relbar} u, v.$$

PROOF. First we consider the configuration formed by a_1, u, v and the basepoint v_*. Over u as basepoint this consists of a, v in Γ_1 with $a \overset{3}{\relbar} v$ and v_* in Γ_2 related to a, v by types 1, 2 respectively. This is certainly realized in Γ.

The configuration a_2uv is also realized in Γ as when we view it as a configuration on the four points v_*a_2uv it is given by Lemma 19.3.41.

So take $u, v \in \Gamma_2$ with $u \overset{1}{\relbar} v$ and $\overset{2}{\sim}$-classes C_1, C_2 containing points a_1, a_2 with the specified types over u, v. If there is some type (i, j) over u, v realized in both C_1 and C_2 we may conclude. Suppose toward a contradiction this is not the case, and consider what types may be realized in C_1 and in C_2, respectively.

By Lemma 19.3.42, if the type (i, j) is realized in one of these classes, then the type (j, i) is realized in the same class. Furthermore the type $(1, 1)$ is not realized in either class.

To begin with we have $(2, 2)$ realized in C_1 and $(1, 3), (3, 1)$ realized in C_2. Thus C_1 realizes neither $(1, 1)$ nor $(3, 1)$ and hence realizes $(2, 1)$ and $(1, 2)$. Then as C_2 does not realize $(2, 1)$ or $(2, 2)$ it realizes $(2, 3)$ and $(3, 2)$. Then similarly C_1 realizes $(3, 3)$ and we have the full list of types realized in each class, which covers all types over (u, v) that may occur for such pairs u, v.

$$\begin{array}{cc} C_1 & C_2 \\ (1,2),\ (2,1),\ (2,2),\ (3,3) & (1,3),\ (3,1),\ (2,3),\ (3,2) \end{array}$$

So for any pair $u, v \in \Gamma_1$ with $u \overset{1}{\relbar} v$ and any $\overset{2}{\sim}$-class C, the set of types realized in C over (u, v) is given by one of these two lists.

Fix a $\overset{2}{\sim}$-class C in Γ_1 and $u, v_1 \in \Gamma_2$ with $u \overset{1}{\relbar} v_1$, such that C realizes the types $(1, 2), (2, 1), (2, 2), (3, 3)$ over u, v_1. In particular $v^1 \cap C \subseteq u^2 \cap C$.

Take $a_1, a_2 \in v_1^1 \cap C$ and $v_2 \in \Gamma_2$ such that

$$v_2 \overset{1}{\relbar} u; \qquad a_1 \in v_2^1 \cap C; \qquad a_2 \in v_2^2 \cap C.$$

Then C realizes the same types over u, v_2 as over u, v_1.

As $a_1 \overset{1}{\relbar} v_1, v_2$ we find $v_1 \overset{2}{\relbar} v_2$ (Lemma 19.3.28). On the other hand $v_1^3 \cap C = v_2^3 \cap C$ since both are equal to $u^3 \cap C$, and this contradicts Lemma 19.3.40. $\qquad\square$

19.3.8. Proposition 19.3.44. Now we can eliminate the locally degenerate imprimitive case.

PROPOSITION 19.3.44. *Suppose that Γ is a homogeneous 3-multi-graph which is locally degenerate imprimitive. Then Γ is of known type.*

PROOF. We make an amalgamation of configurations $a1234$ and $b1234$ with the type structure as shown in Figure 1.

	b	1	2	3	4
a	?	1	2	1	2
1	1		1	2	2
2	2	1		2	2
3	1	2	2		1
4	3	2	2	1	

FIGURE 1. Amalgamation for Proposition 19.3.44.

In this diagram, 1a2b forces $a \overset{2}{-} b$ and then 3ab4 gives a contradiction. So it suffices to show that the factors $a1234$ and $b1234$ are realized in Γ.

In the configuration $a1234$, taking a as basepoint gives the configuration covered by Lemma 19.3.41.

In the configuration $b1234$, taking 1 as basepoint gives the configuration covered by Lemma 19.3.43.

This completes the proof. □

19.4. 3-multi-graphs: conclusion

We review what was already said at the beginning of our analysis, concerning the stage now reached in the classification of homogeneous 3-multi-graphs.

Combining the results of Amato, Cherlin, and Macpherson [2021] with the analysis in this chapter, we see that we have a satisfactorily systematic catalog of the homogeneous 3-multi-graphs, one which has some prospects for being complete. Namely, going forward, we may suppose that our homogeneous 3-multi-graph is primitive, and that any forbidden triangle is monochromatic, while at least one monochromatic triangle type is realized.

Since the target in this case is to identify the graph as the Fraïssé limit of a free amalgamation class, this brings us at last to the "generic" case. We remark that the treatment of Amato, Cherlin, and Macpherson [2021] includes cases where the class has free amalgamation (with value 2, if the 2-types are identified with distances 1,2,3), so that we have already crossed over the threshold of the generic case.

Here one would expect that one would soon come to rely heavily on Lachlan's Ramsey theoretic approach, which goes along with a change of categories, where one replaces the original 3-multi-graph by a more structured object consisting of the realizations of two 1-types over a point, which is helpful in formulating matters in inductive terms. So in order to apply this method, one of the first questions is whether one has sufficient information about these 1-types.

In the favorable case in which there is a 2-type p such that some finite p-clique is forbidden, one will focus on Γ_p, which in principle can be identified by induction, but only after eliminating various special cases. We have just eliminated one of these, namely the locally degenerate imprimitive case: this gives part of the analysis when a monochromatic triangle is forbidden. Thus while this falls on the "generic" side of the analysis, still it is only concerned with the elimination of further sporadic examples in which Γ_p does not have the expected structure.

In particular, when Γ omits a monochromatic triangle of type p, we may suppose that Γ_p is primitive. As it involves only two non-trivial 2-types, we view it as a homogeneous graph. Then Γ_p may be either a Henson graph or a random graph up to a choice of notation, since graph complementation simply amounts to relabeling the 2-types.

If Γ_p is a Henson graph, then one knows in particular which type is available to be used for free amalgamation. If Γ_p is merely a random graph, then even this point is not clear at this stage of the analysis. So this last case may enter new territory.

Once those special cases are also disposed of, one will suppose that all triangles are realized. This then gives very little control over the constraints on the class. The target would be to show in this case that one of the 2-types does not occur in any constraint. If there is any forbidden clique (monochromatic constraint) then one has leverage for an inductive argument. Otherwise we must somehow show that all minimal constraints involve just two 2-types if taken together. This was accomplished by surprisingly direct methods in Amato, Cherlin, and Macpherson [2021], with very particular use of the triangle inequality (i.e., a particular constraint involving two 2-types already). It remains to be seen whether the germ of something more general can be found in that analysis. Namely, if one has a minimal constraint involving only two 2-types, one would like to use it to show that all minimal contraints involve the same 2-types. Finally, there is the case—which should not occur—in which all minimal constraints involve all 2-types, and this remains mysterious.

Chapter 20

IMPRIMITIVE HOMOGENEOUS
2-MULTI-TOURNAMENTS

20.1. Introduction

Generalizing the problem taken up in Part I, we ask for the classification of
the homogeneous 2-multi-tournaments. By definition, these are tournaments
together with a coloring of the arcs by two colors: that is, there are two pairs
of anti-symmetric quantifier-free 2-types: if we view the colors and the arcs
separately, 2-multi-tournaments may also be thought of as a set equipped with
a graph relation and a tournament relation, so that linearly ordered graphs are
a very special case.

This turns out to be somewhat more complex than we would have expected,
based on prior results—certainly more troublesome than the case of 3-multi-
graphs, which we find surprising. Perhaps one can justify it after the fact
by the observation that while the classification of homogeneous structures in
a language with finitely many equivalence relations and the classification of
homogeneous structures in a language with finitely many linear orders can both
be carried out completely, the latter lies considerably deeper than the former.
That is, in the transitive case, anti-symmetry is much harder to deal with
than symmetry. The reason we find this surprising is that the homogeneous
directed graphs consist largely of the natural analogs of homogeneous graphs,
so that de-symmetrizing just one of the two 2-types does not radically alter the
statement of the result (though it does complicate the proof).

This chapter and the next two run somewhat parallel to the discussion of
homogeneous 3-multi-graphs in Chapter 19, but the discussion has to begin at
an earlier stage in the analysis, and eventually becomes more difficult.

We begin with a catalog of the known examples. Then we will give a full
classification of the *imprimitive* homogeneous 2-multi-tournaments. This had
not previously been worked out, but the resulting classification turns out to be
very similar to the classification of imprimitive homogeneous directed graphs.
In a systematic study of homogeneous 2-multi-tournaments this topic would
most naturally have preceded the study in Part 1: imprimitivity is the most

special case of a triangle constraint. In the case of homogeneous 3-multi-graphs the imprimitive case had been treated previously, and we were able to begin the discussion in the previous chapter at a more advanced point.

Once that point is disposed of, the general analysis of triangle constraints for homogeneous 2-multi-tournaments will occupy the following chapters. As we will see, our analysis will leave more to be done on that question.

We fix notation for the irreflexive 2-types as follows: these will be denoted by $\overset{1}{\to}, \overset{2}{\to}$, together with their reversals by $\overset{1}{\leftarrow}, \overset{2}{\leftarrow}$.

Notation 20.1.1.

1. Much as in Chapter 19, we use the notation a^i for

$$\{x \in \Gamma \mid a \overset{i}{\to} x\},$$

and the analogous notation a^p where p is $\overset{i}{\to}$.

However here we have the four 2-types $1, 2, 1^{\text{op}}, 2^{\text{op}}$ where i represents the type $\overset{i}{\to}$ and i^{op} represents the type $\overset{i}{\leftarrow}$. We generally prefer the notation $a^{i^{\text{op}}}$ to the equally reasonable notation $^i a$; the main exception will be noted next.

When we are less concerned with the color of the arc (1 or 2) than with its direction, we sometimes find the following notation more convenient.

2. For Γ a 2-multi-tournament with arc relations $\overset{1}{\to}, \overset{2}{\to}$, we let $\to$ be the arc relation $\overset{1}{\to} \cup \overset{2}{\to}$ (the underlying tournament). For $a \in \Gamma$ we set

$$a' = \{x \in \Gamma \mid a \to x\}; \qquad 'a = \{x \in \Gamma \mid x \to a\}.$$

We mention in passing that another notation similar in appearance will be used with a very different meaning later on, namely the notation T^i, for T a finite tournament with arc relation $\to$ (Notation 21.1.3). But the latter notation occurs in a very different context and confusion seems unlikely.

20.2. A catalog of homogeneous 2-multi-tournaments

We first list all the known homogeneous 2-multi-tournaments, up to choice of language (labeling of 2-types), under the following headings: degenerate, imprimitive, finite—and the rest. Here we exclude from each category any example already considered under one of the previous headings. There is some technical terminology to be explained afterward. But we leave the detailed discussion of the final group of five unusual 3-constrained examples for the following chapter.

A far more technical presentation of the catalog is tabulated in Tables 22.1 and 22.2, pp. 219, 220, where the structures are given in terms of the constraints defining the corresponding amalgamation classes.

(I) **Degenerate** (omitting some 2-type):

– a homogeneous tournament in the language $\overset{1}{\to}$ or $\overset{2}{\to}$.

(II) **Imprimitive, non-degenerate**
- Compositions $T_1[T_2]$ with T_i a homogeneous tournament with arc relation $\overset{i}{\to}$.
- Shuffled of type $\mathbb{Q}$ or $\mathbb{S}$.
- Semi-generic of type $\mathbb{Q}$, $\mathbb{S}$, or T^∞.
- Generic de-symmetrization of an imprimitive directed graph of type $n * T$ with T an infinite homogeneous tournament and $2 \leq n \leq \infty$.

(III) **Finite, primitive, non-degenerate**: 5 vertices
- an oriented pentagram.

(IV) **Infinite, primitive, non-degenerate**
- A generic linear extension of a partial order.
- The free join of two homogeneous primitive infinite structures Γ_1, Γ_2, where
  - Γ_1 is a tournament and
  - Γ_2 is either a homogeneous tournament or a homogeneous graph.
- A generic 2-splitting of a homogeneous tournament with strong amalgamation, with forbidden tournaments of one edge type.
- A generic de-symmetrization of a primitive homogeneous directed graph with strong amalgamation.
- 3-constrained, and not previously listed.
  - (a) $\overset{1}{\to}$ is a partial order and $a \overset{1}{\to} b \overset{1}{\to} c \overset{2}{\leftarrow} a$ is forbidden.
  - (b) Forbid $a \overset{1}{\to} b \overset{1}{\to} c \overset{1}{\to} a$ and $a \overset{1}{\to} b \overset{1}{\to} c \overset{2}{\to} a$, that is forbid 3-cycles with two edges of type $\overset{1}{\to}$.
  - (c) Forbid $a \overset{1}{\to} b \overset{1}{\to} c \overset{1}{\leftarrow} a$ (L_3) and $a \overset{1}{\to} b \overset{1}{\to} c \overset{2}{\to} a$ (C_3).
  - (d) Forbid $a \overset{1}{\to} b \overset{1}{\to} c \overset{1}{\leftarrow} a$, $a \overset{1}{\to} b \overset{1}{\to} c \overset{1}{\to} a$ (L_3, C_3) and $a \overset{1}{\to} b \overset{1}{\to} c \overset{2}{\to} a$.
  - (e) If (a, b) and (a, c) realize the same type, then $b \overset{1}{\to} c$ is forbidden; and the 3-cycles $a \overset{1}{\to} b \overset{1}{\to} c \overset{1}{\to} a$, $a \overset{2}{\to} b \overset{2}{\to} c \overset{2}{\to} c$ are forbidden. In particular, for each $a \in \Gamma$ and each 1-type p over a, $(a^p, \overset{2}{\to})$ is a linear order.
- A variant of (d) forbidding L_n^1 with $n > 3$.

Since we are about to launch into a detailed study of the imprimitive case, we will first discuss the various technical notions which come into the specification of the known examples in the primitive cases, and then address the imprimitive case in the following section.

In the *degenerate* case we assume that not all 2-types are realized, so we may suppose that the language reduces to $\overset{1}{\to}$, and we are considering the five homogeneous tournaments, which are denoted as follows.

(a) I (trivial, one element).

(b) The 3-cycle C_3.

(c) $\mathbb{Q}$ the rational order, as a tournament.

(d) The generic local order $\mathbb{S}$.

(e) The generic (or homogeneous universal, or random) tournament T^∞.

DEFINITION 20.2.1. The *oriented pentagram* is the 2-multi-tournament on $\mathbb{Z}/5\mathbb{Z}$ with arc relations $x \xrightarrow{i} y$ for $i = 1$ or 2 given by

$$y - x = i.$$

We now discuss a number of "generic constructions" involving strong amalgamation classes. The homogeneous structures which correspond to strong amalgamation classes are those in which every set is algebraically closed in the model theoretic sense; in particular, if the structure in question is imprimitive then the corresponding equivalence classes are infinite, and in practice (when the homogeneous structures are known) this condition tends to be sufficient as well.

DEFINITION 20.2.2 (Generic constructions). Let $\mathcal{A}_1, \mathcal{A}_2$ be two strong amalgamation classes in languages L_1, L_2, which we assume are disjoint (or which we make disjoint, if necessary).

1. Their *free join* $\mathcal{A} = \mathcal{A}_1 * \mathcal{A}_2$ is the class of finite structures in the language $L = L_1 \cup L_2$ whose reducts to L_i are in $\mathcal{A}_i$ for $i = 1, 2$.

This is then a strong amalgamation class. (We need to assume $\mathcal{A}_1, \mathcal{A}_2$ are strong amalgamation classes even to deduce that their free join is an amalgamation class.)

This terminology may be transferred to the Fraïssé limits $\Gamma_1, \Gamma_2, \Gamma$ of $\mathcal{A}_1, \mathcal{A}_2, \mathcal{A}$: we may write

$$\Gamma = \Gamma_1 * \Gamma_2$$

and call this structure the free join.

2. In a frequently occurring case, $\mathcal{A}_2$ is the class of finite linear orders, and then $\Gamma_1 * \mathbb{Q}$ is Γ_1 expanded by a *generic linear order*.

3. Similarly, if Γ is the Fraïssé limit of a strong amalgamation class $\mathcal{A}$ and p is a non-trivial 2-type of Γ, named by the language, to *split* the 2-type p *generically*, we replace the name for p by a set of names P, and replace the amalgamation class $\mathcal{A}$ by the structures in the altered language in which each occurrence of p is replaced by an arbitrary symbol from P.

The following case is of interest here. If we split an anti-symmetric 2-type into two anti-symmetric 2-types, an infinite homogeneous tournament gives rise to a homogeneous 2-multi-tournament, with the arc relation *generically 2-split*.

4. When we split a 2-type into a set of 2-types P, we may also single out a proper subset P_0 of P closed under permutation of variables, and then forbid in

addition some set of P_0-structures, using types in $P \setminus P_0$ to complete amalgams where needed. That is, whenever the original type p would have been used to complete an amalgam, types in $P \setminus P_0$ may be used instead.

The additional constraints play the role of Henson constraints: additional constraints compatible with the chosen amalgamation procedure.

5. Another way to split a type generically is to reduce its symmetry group.

The case of interest here is the following. Take a homogeneous directed graph Γ associated with a strong amalgamation class $\mathcal{A}$ (in other words, if the directed graph is imprimitive, we require the equivalence classes to be infinite). We split the symmetric "non-edge" relation into a pair of anti-symmetric 2-types. In terms of amalgamation classes, we replace the directed graphs in $\mathcal{A}$ by 2-multi-tournaments by orienting the non-edges in all possible ways.

The Fraïssé limit of the resulting class is the *generic de-symmetrization* of Γ: it is a homogeneous 2-multi-tournament.

The *generic linear extension of a partial order* has a more specialized character. It is the Fraïssé limit of the class of all finite structures in a language with a symbol for a partial order $\preceq$ and a symbol for a linear order $\leq$, under the requirement the the linear order extends the partial order (this example is familiar from structural Ramsey theory and falls into the framework of Part I of this work, as an exceptional case known from prior work).

One must pay a little attention to the correct amalgamation procedure. Given an amalgamation diagram, the union of the partial orders on the factors gives an acyclic digraph, and the union of the linear orders gives an acyclic digraph; in amalgamating, one should take the minimal extension of the former to a partial order, and any extension of the latter to a total order, so as to meet the required constraint.

Now we have dealt with the terminology and notation used in the catalog, with the exception of the aforementioned cases, namely: (1) the imprimitive 2-multi-tournaments, which will be our next topic, and (2) the five exceptional 3-constrained 2-multi-tournaments which will occupy us in the following chapter. The latter are explicitly defined, but it is less clear at this point in our discussion that the definitions give amalgamation classes, and we have no satisfactory conceptual framework for them.

Now taking up the classification problem (completeness of the catalog) from the very beginning, the first point is the following.

FACT 20.2.3 (Lachlan [1984], [1986]). *The degenerate or finite homogeneous 2-multi-tournaments are as specified in the catalog above.*

In particular, the finite homogeneous 2-multi-tournaments are the compositions of finite tournaments (so in the non-degenerate case, $C_3[C_3]$) together with the oriented pentagram.

In Lachlan [1986] our 2-multi-tournaments are called 2-*tournaments*, and the homogeneous ones are also referred to as the *members of* $\mathrm{Hom}(2; 1, 2; \omega)$, meaning: binary homogenous, with only 1 symmetric 2-type (equality) and with 2 pairs of asymmetric 2-types; in this notation, the label ω signifies that the structures should be stable (which for the purposes of classification, given the existing theory, is not much more general than requiring finiteness).

With these cases out of the way, we come next to the imprimitive case. In this case we will show that the catalog is complete.

20.3. Imprimitive homogeneous 2-multi-tournaments

DEFINITION 20.3.1. The minimal symmetric and reflexive extension of the relation $\overset{i}{\to}$ is denoted $\overset{i}{\sim}$. This relation is of interest mainly when it is an equivalence relation.

In dealing with imprimitive homogeneous 2-multi-tournaments we may suppose that $\overset{1}{\sim}$ is an equivalence relation.

We will prove the following, along the lines of the classification of the imprimitive homogeneous directed graphs given in Cherlin [1987].

PROPOSITION 20.3.2. *Let Γ be an imprimitive homogeneous 2-multi-tournament. Then up to a change of language Γ is one of the following structures.*

(a) *A composition $T_2[T_1]$ with T_i an $\overset{i}{\to}$-tournament for $i = 1, 2$.*

(b) *(Shuffled type) Γ is derived from a homogeneous local order $T \cong \mathbb{Q}$ or $\mathbb{S}$ with arc relation $\to$ which is partitioned into n dense pieces, with $2 \leq n \leq \infty$. If $a \to b$ holds in T then we take $a \overset{1}{\to} b$ if a, b are in the same piece of T, and $a \overset{2}{\to} b$ if a, b are in distinct pieces of T.*

(c) *(Semi-generic type) Generic imprimitive with infinite components (type $\mathbb{Q}$, $\mathbb{S}$, or T^∞) and satisfying the* parity constraint *described below.*

(d) *(Generic type) The generic de-symmetrization of an imprimitive homogeneous directed graph $n \cdot T$ with $2 \leq n \leq \infty$, where T is an infinite homogeneous tournament. Here $n \cdot T$ denotes the disjoint union of n copies of T and the generic de-symmetrization is denoted $n * T$.*

There is a lot of notation to address here, and we should also take a moment where necessary to check the homogeneity of the structures listed in our catalog, once they have been properly defined (Lemma 20.3.5 below).

We first describe the more direct constructions.

DEFINITION 20.3.3.

1. The *composition $T_2[T_1]$* of two tournaments is the result of replacing each point of T_2 by a copy of T_1, and taking $\overset{1}{\to}$ to be the tournament relation on T_1 within each component, and $\overset{2}{\to}$ to be the tournament relation on T_2, between the components.

2. If T is an infinite homogeneous local order (the rational order $\mathbb{Q}$ or the generic local order $\mathbb{S}$) then the *shuffled* 2-multi-tournament of type T with n classes is formed by dividing T into n dense subsets and representing the tournament relation in T by $\xrightarrow{1}$ within the subsets and by $\xrightarrow{2}$ between them.

We will denote this shuffled 2-multi-tournament by $T^{(n)}$: thus we have $\mathbb{Q}^{(n)}$ and $\mathbb{S}^{(n)}$.

The term *shuffled* is perhaps more apt in the case of $\mathbb{Q}^{(n)}$ than in the case of $\mathbb{S}^{(n)}$.

3. Generic de-symmetrization has been defined in Definition 20.2.2. It applies to any homogeneous directed graph associated with a strong amalgamation class, and produces a homogeneous 2-multi-tournament with the non-edge relation split into a pair of anti-symmetric 2-types.

In the imprimitive case we begin with an infinite homogeneous tournament T, form the disjoint sum of n copies of T, which we denote by

$$n \cdot T,$$

and then take the generic de-symmetrization, which we denote by

$$n * T.$$

Thus this is the disjoint union of n copies of T with arc relation $\xrightarrow{1}$, and the arc relation $\xrightarrow{2}$ between these copies is imposed generically, via the Fraïssé theory.

We also have the "semi-generic" variation on $\infty * T$ to discuss.

DEFINITION 20.3.4 (Semi-generic 2-multi-tournament). We define the *semi-generic* 2-multi-tournament with components of specified type T (an infinite homogeneous tournament) via the Fraïssé theory, by the following constraints.

(a) The connected components for the relation $\xrightarrow{1}$ are of type T;
(b) *The parity constraint:* between two pairs $a_1 \xrightarrow{1} a_2$ and $b_1 \xrightarrow{1} b_2$, there are an even number of arcs $a \xrightarrow{2} b$ with $a = a_1$ or a_2 and $b = b_1$ or b_2.

LEMMA 20.3.5. *The 2-multi-tournaments listed in Proposition* 20.3.2 *are all homogeneous.*

PROOF. Composition of homogeneous structures in disjoint languages preserves homogeneity.

For the case of the shuffled 2-multi-tournaments $\mathbb{Q}^{(n)}$ and $\mathbb{S}^{(n)}$, one needs to know that a homogeneous local order remains homogeneous in an expanded language giving a partition into dense subsets.

Skolem observed in Skolem [1920, §4, Sätze 2 & 3] that the proof of $\aleph_0$-categoricity for countable dense linear orders (Cantor) without endpoints extends to the case of expansions of such a linear order by a fixed number—possibly infinite—of dense subsets. He applied the back and forth method as in Hausdorff's exposition, which Skolem cites; this is a more flexible and transparent method than Cantor's original one. Exactly the same argument gives homogeneity as well.

It follows that the shuffled 2-multi-tournaments are homogeneous in the language expanded by the corresponding unary predicates. To see that they are homogeneous as 2-multi-tournaments requires checking in addition that the equivalence classes can be freely permuted by automorphisms of the structure. This is a consequence of the uniqueness up to isomorphism, since permuting the classes gives another structure of the same type.

The construction in the semi-generic case is handled by direct verification of the amalgamation property. It suffices to deal with 2-point amalgamation problems $A \cup \{a_1\}$, $A \cup \{a_2\}$ over A.

Suppose first that there are points b_1, b_2 in A in the same $\xrightarrow{1}$-components as a_1, a_2. Then there are two cases: either b_1, b_2 lie in the same $\xrightarrow{1}$-component and we amalgamate as in this component (with no identification of points), or they lie in distinct components and the parity constraint dictates the amalgam. In these cases, one must check that the parity constraint is preserved.

If there is no such pair of points b_1, b_2 in A then one takes $a_1 \xrightarrow{2} a_2$ or the reverse, as one likes, and there is nothing left to be checked.

In the last case (generic imprimitive) we begin with a directed graph $n \cdot T$ consisting of n unrelated copies of the infinite homogeneous tournament T, where $2 \leq n \leq \infty$. One notes that T corresponds to a strong amalgamation class, and thus $n \cdot T$ also corresponds to a strong amalgamation class, so that generic de-symmetrization is possible, with n held fixed. One may then take the Fraïssé limit.   □

As one might expect, the proof of the classification stated in Proposition 20.3.2 is lengthy, as it deals separately with the various cases arising. We fix the following notation throughout the discussion.

Notation 20.3.6. Let Γ be a homogeneous imprimitive 2-multi-tournament. We may suppose that $\overset{1}{\sim}$ gives an equivalence relation on Γ. The $\overset{1}{\sim}$-classes, with their induced structure, will be called the *components* of Γ.

Our standing hypothesis will be that Γ is not a composition, but we take note of this hypothesis explicitly as we proceed. Let us agree to consider the degenerate case as a composition: $T = I[T]$ where I has no arcs but the language nonetheless contains a symbol for one of the two arc relations. That is, we assume nondegeneracy as well, but are not obliged to mention it separately.

Lemma 20.3.7. *Let Γ be a homogeneous imprimitive 2-multi-tournament, and not a composition. Then the components of Γ are infinite, namely: one of the two local orders $\mathbb{Q}$, $\mathbb{S}$, or the generic tournament T^∞.*

PROOF. By the classification of homogeneous tournaments, the alternative is that the components of Γ are 3-cycles. If C is one such, and $a \in \Gamma \setminus C$, then $a' \cap C$ and $'a \cap C$ are homogeneous tournaments. So one of them is empty, and Γ is a composition.   □

A point to be kept in mind, as it tends to be invoked without explicit mention, is the following.

LEMMA 20.3.8. *Let Γ be a homogeneous imprimitive 2-multi-tournament, and not a composition. Let C_1, C_2 be two components, and $a \notin C_1$, $b \notin C_2$. Then there is an automorphism Γ carrying (b, C_2) to (a, C_1).*

PROOF. It suffices to treat the case in which we make a particular choice for a, C_1, leaving b, C_2 to vary.

So as Γ is not a composition, we choose a, C_1 so that a realizes both possible types $\xrightarrow{2}$ and $\xleftarrow{2}$ over C_1. Then we fix b, C_2, and $c \in C_2$, and carry (b, c) to (a, d) for a suitable choice of $d \in C_1$.   □

In dealing with local orders we generalize some notions associated with linear orders. We have already had occasion to refer to density. We also need the notion of convex set, and the related notion of Dedekind cut.

There is more than one notion of convexity available in the context of tournaments (notably, there is a notion of *interval* stricter than the definition we use).

DEFINITION 20.3.9. Let S be a tournament and I a subset. Then I is *convex* if there is no triple a, b, c with $a \rightarrow b \rightarrow c$, $a \rightarrow c$, and $a, c \in I$, $b \notin I$.

LEMMA 20.3.10. *Let S be a local order, $I \subseteq S$ convex and proper. Then I is linearly ordered by the arc relation.*

PROOF. Suppose there is a 3-cycle (a, b, c) in I. Then S is the union of the intervals $[a, b]$, $[b, c]$, $[c, a]$ defined by

$$a \rightarrow x \rightarrow b \qquad\qquad b \rightarrow x \rightarrow c \qquad\qquad c \rightarrow x \rightarrow a$$

and each of these is contained in I, so $I = S$.   □

We remark also that for S a local order and $a \in S$, the sets a' and $'a$ are convex.

DEFINITION 20.3.11. Let S be a local order. A *Dedekind cut* in S is an unordered pair of non-empty convex subsets whose union is S.

The Dedekind cut is *proper* if neither part has an endpoint.

In the linear case the ordering itself distinguishes a "left" and "right" side in the Dedekind cut. In the non-linear case a Dedekind cut corresponds to two antipodal points in a natural completion, after making appropriate conventions for cuts associated to existing elements. In the proper case, neither of these points lies in the original structure. In the case of $\mathbb{S}$, the Dedekind completion may be identified with the unit circle and $\mathbb{S}$ itself with any countable dense subset of the circle containing no antipodal pairs of points, and with an arc relation derived from an orientation of the circle. On the Dedekind completion of $\mathbb{S}$ the arc relation gives a directed graph rather than a tournament; antipodal pairs are not related by an arc.

The following generalizes the comparability of Dedekind cuts in the linearly ordered case.

LEMMA 20.3.12. *Let S be a local order with two decompositions into proper convex parts $S = S_1 \cup S_2$ and $S = S_1' \cap S_2'$. Let $i, j \in \{1, 2\}$. Then the following hold.*

(a) $S_i \cap S_j'$ *is a terminal or initial segment of S_i (possibly empty, or all of S_i).*

(b) *If $S = \mathbb{S}$ is the generic local order, the sets S_1, S_2 have no endpoints, and $\{S_1, S_2\} \neq \{S_1', S_2'\}$, then the segments $S_i \cap S_j'$ are proper subsets of S_i and S_j.*

PROOF. We may suppose $i = j = 1$.

For the first point, if we suppose otherwise then it follows easily that $S_1 \cap S_1'$ is bounded below and above in S_1, say by a, b respectively. Then $a, b \in S_2$ and by convexity $S_1 \cap S_1' \subseteq S_2$, so $S_1 \cap S_1' = \emptyset$.

For the second point, in the contrary case we may suppose $S_1 \subseteq S_1'$, and that S_1 is a proper initial segment of S_1'. Since S_1' has no endpoint, we may take $a \to b$ in $S_1' \setminus S_1$ and extend to a 3-cycle (a, b, c). Then $c \in S_2' \subseteq S_2$ so the 3-cycle (a, b, c) lies in S_2, a contradiction.   □

20.3.1. The shuffled case.

DEFINITION 20.3.13. Let Γ be a homogeneous imprimitive 2-multi-tournament.

1. Γ is said to be of *shuffled type* if the following hold.
   - The components are dense local orders.
   - For each component C and $a \notin C$, the two sets $a' \cap C$ and $'a \cap C$ are convex and non-trivial (hence also proper subsets).
2. Γ is said to be of *general type* if its components are infinite, and for each component C and $a \notin C$, we have $a' \cap C, 'a \cap C$ nonempty. Furthermore, if the components are local orders then we require $a' \cap C$ and $'a \cap C$ to be dense in C.

LEMMA 20.3.14. *Let Γ be a homogeneous imprimitive 2-multi-tournament, not a composition, with components of type $\mathbb{Q}$ or $\mathbb{S}$. Then Γ is either of shuffled type, or of general type.*

PROOF. Fix a component C and $a \notin C$.

By Lemma 20.3.8, $a' \cap C$ and $'a \cap C$ are both nonempty. So we may suppose now that the components are local orders of type $\mathbb{Q}$ or $\mathbb{S}$.

CLAIM 1. *Either $a' \cap C$ is convex, or $'a \cap C$ is dense in C.*

We suppose $a' \cap C$ is not convex and fix $b_1, b_2, c \in a' \cap C$ with $b_1 \to b_2$, $b_1 \to c \to b_2$, $c \to a$. We show $'a \cap C$ is dense in C. With the exception of the point a, we now work inside C.

So fix $x_1, x_2 \in C$ with $x_1 \to x_2$ and consider the corresponding interval (x_1, x_2) in C. We look for $y \in (x_1, x_2)$ with $y \xrightarrow{2} a$.

Take $y_1 \to y_2$ in the interval (x_1, x_2). We may suppose $a \xrightarrow{2} y_1, y_2$, as otherwise we are done. By homogeneity some automorphism α carries a, b_1, b_2 to a, y_1, y_2, and then $\alpha(c)$ lies in (y_1, y_2). This proves the claim.

In particular, if $a' \cap C$ is not convex, then since $'a \cap C$ is a proper subset of C, it is not convex either, and so by the dual of Claim 1 $a' \cap C$ is also dense.

Thus $a' \cap C$ is either convex or dense, and $'a \cap C$ is either convex or dense, and thus either both of these sets are convex, or both are dense.

By Lemma 20.3.8 the same applies uniformly to all pairs (a, C). Thus Γ is either shuffled or of general type. $\qquad\square$

In the present subsection we consider the shuffled case. According as the components are of type $\mathbb{Q}$ or $\mathbb{S}$, we distinguish two cases.

(a) Type $\mathbb{Q}$: for each component C and $a \notin C$, a determines a Dedekind cut in $\mathbb{Q}$, consisting of an initial segment and a terminal segment (not realized in C).

(b) Type $\mathbb{S}$: for each component C and $a \notin C$, a determines a partition of $\mathbb{S}$ into two linearly ordered convex subsets, without endpoints (a Dedekind cut, in the terminology of Definition 20.3.11).

In the case of type $\mathbb{Q}$, we would like to know which of the two parts of the Dedekind cut corresponds to $a' \cap C$; up to a change of language—specifically, we may reverse the relation $\xrightarrow{1}$—we may suppose that $a' \cap C$ is a terminal segment of C, and if this holds for one such pair a, C it will hold for all. In the case of type $\mathbb{S}$, the situation is more symmetrical at the outset.

DEFINITION 20.3.15. Let Γ be an imprimitive homogeneous 2-multi-tournament of shuffled type. Then Γ will be said to be *properly shuffled* if one of the following conditions applies.

(a) The components are of type $\mathbb{Q}$, and for C a component, $a \notin C$, the set $a' \cap C$ is a terminal segment of C.

(b) The components are of type $\mathbb{S}$, and for C a component, $a, b \notin C$, if $a \xrightarrow{1} b$ then $b' \cap C$ meets $a' \cap C$ in a terminal segment.

LEMMA 20.3.16. *Let Γ be an imprimitive homogeneous 2-multi-tournament of properly shuffled type, with components of type $\mathbb{Q}$. Then the relation $\to$ defined as $\xrightarrow{1} \cup \xrightarrow{2}$ is transitive.*

PROOF. We consider potential 3-cycles $a \to b \to c \to a$ according to the number of $\overset{1}{\sim}$-classes involved.

On each class, $\to$ coincides with $\overset{1}{\to}$, and is transitive.

On the union of two classes $C_1 \cup C_2$, we may suppose $a \in C_1$, $b, c \in C_2$, and the definition of proper shuffling applies.

Finally, suppose we have a 3-cycle

$$a_1 \to a_2 \to a_3 \to a_1$$

with a_1, a_2, a_3 in distinct $\overset{1}{\sim}$-classes C_1, C_2, C_3. Then we take $b_2 \in C_2$, $b_3 \in C_3$ satisfying

$$a_1 \to b_3 \to b_2$$

(Figure 2).

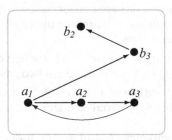

FIGURE 2. Transitivity.

By transitivity of the relation $\to$ on $C_1 \cup C_3$, $C_2 \cup C_3$, and finally $C_1 \cup C_2$, we find, successively,

$$a_3 \to b_3; \qquad a_2 \to b_3; \qquad a_2 \to b_2; \qquad a_1 \to b_2.$$

Now take $u \in C_3$ with

$$u \to a_1, a_2, b_2$$

(these conditions define an initial segment of C_3).

By homogeneity, there is an automorphism α carrying (a_1, a_2, u) to (a_1, b_2, u). Set $c_3 = \alpha(a_3)$. Then

$$a_1 \to b_2 \to c_3 \to a_1.$$

By transitivity on $C_2 \cup C_3$ and $C_1 \cup C_3$ we find successively

$$b_3 \to c_3, \qquad\qquad b_3 \to a_1,$$

a contradiction. □

Lemma 20.3.17. *Let Γ be an imprimitive homogeneous 2-multi-tournament of shuffled type, with components of type $\mathbb{Q}$, and let $n = |\Gamma/\overset{1}{\sim}|$. Then up to a change of language $\Gamma \cong \mathbb{Q}^{(n)}$, the generic shuffled 2-multi-tournament of type $\mathbb{Q}$ with n components.*

Proof. As we have remarked, up to a change of language Γ is properly shuffled. So we assume that this choice of language has been made.

The relation $\rightarrow$ defined as $\overset{1}{\rightarrow} \cup \overset{2}{\rightarrow}$ gives a linear order on Γ. It suffices to show that the $\overset{1}{\sim}$-classes are dense in Γ.

Take three components C_1, C_2, C_3, not necessarily distinct, and $u_1 \in C_1$, $u_2 \in C_2$. We show the interval (u_1, u_2) meets C_3.

Take $a \in C_3$, $v_1 \in C_1$, $v_2 \in C_2$, with $v_1 \rightarrow a \rightarrow v_2$. Take $b \in C_3$ with $b \rightarrow a, u_1, v_1, u_2, v_2$. There is an automorphism α taking (b, v_1, v_2) to (b, u_1, u_2) and hence taking C_1, C_2, C_3 to themselves. Then $\alpha(a)$ lies in the desired interval.

As we mentioned previously, Skolem showed that there is a unique structure of this type, in the language including unary predicates for the components, so the reduct we are considering must be the shuffled 2-multi-tournament $\mathbb{Q}^{(n)}$. □

Now we turn to 2-multi-tournaments of shuffled type with components of type $\mathbb{S}$. We deal first with the properly shuffled cases, along similar lines to the foregoing.

LEMMA 20.3.18. *Let Γ be an imprimitive homogeneous 2-multi-tournament of shuffled type, with n components of type $\mathbb{S}$, $n \geq 3$. Then for $u \in \Gamma$, u^2 is an imprimitive homogeneous 2-multi-tournament of shuffled type, with $(n-1)$ components of type $\mathbb{Q}$.*

We note that for $n = 2$, the notion of shuffled type would not make sense for a tournament with $(n-1)$ components; hence the restriction $n \geq 3$.

PROOF. By hypothesis the components of u^2 are of type $\mathbb{Q}$ and by definition, for $v \in u^2$, the sets v^2 and $v^{2^{op}}$ meet each component of u^2 in a convex subset. What needs to be checked is that u^2 is not itself a composition.

Suppose on the contrary that u^2 is a composition. Fix a $\overset{1}{\sim}$-class C of Γ.

CLAIM 1. *The elements of C realize only two types over $\Gamma \setminus C$.*

Take $\overset{1}{\sim}$-classes C_1, C_2 of Γ with C, C_1, C_2 distinct and $u \in C_2$. As u^2 is a composition, the types of the triples (u, C, C_1) and (u, C_1, C) are distinct. Let $v \in u^2 \cap C$. Then v^2 does not split $u^2 \cap C_1$. If v^2 splits $u^{2^{op}} \cap C_1$, then we may take a vertex $w \in u^{2^{op}} \cap C_1$ and vertices v', w' with $v' \in u^2 \cap C_1$, $w' \in u^{2^{op}} \cap C$ so that (v, w) and (v', w') have the same type, and hence (u, v, w) and (u, v', w') have the same type. But then (u, C, C_1) and (u, C_1, C) have the same type, a contradiction. Hence $v^2 \cap C_1$ and $v^{2^{op}} \cap C_1$ must be coincide in some order with $u^2 \cap C_1$ and $u^{2^{op}} \cap C_1$; furthermore for all $v \in u^2 \cap C$, the order is the same. Thus for $v_1, v_2 \in u^2 \cap C$, and any $\overset{1}{\sim}$-class C_1 other than C or C_2, the vertices v_1 and v_2 realize the same type over C_1. Replacing C, C_1, C_2 by C, C_2, C_1 gives the same conclusion for C_2.

This proves the claim.

The relation defined on Γ by "x, y lie in the same $\overset{1}{\sim}$-class C, and realize the same type over $\Gamma \setminus C$" is invariant under Aut Γ and thus definable. Therefore the same relation restricted to a single class C is definable in the language of C. But as C is primitive, this is a contradiction.   □

LEMMA 20.3.19. *Let Γ be an imprimitive homogeneous 2-multi-tournament of properly shuffled type, with n components of type $\mathbb{S}$. Then*

$$\Gamma \cong S^{(n)}$$

the generic imprimitive 2-multi-tournament with the specified parameters.

PROOF. Given two $\overset{1}{\sim}$-classes C_1, C_2, let $\hat{C}_1$, $\hat{C}_2$ be their Dedekind completions (a point may be thought of as given locally by a Dedekind cut in some convex linearly ordered part).

We define $f_{C_1,C_2} : C_1 \to \hat{C}_2$ by $f_{C_1,C_2}(a) = \min(a' \cap C_2)$. By the definition of proper shuffling, this gives an isomorphic embedding of C_1 into $\hat{C}_2$. The image is dense, so we may use the same notation for the extension $f_{C_1,C_2} : \hat{C}_1 \cong \hat{C}_2$. Composition $f_{C_2,C_3} \circ f_{C_1,C_2}$ gives an isomorphism from $\hat{C}_1$ to $\hat{C}_3$. If this is distinct from f_{C_1,C_3} then we get an automorphism of $\hat{C}_1$ and in particular a nontrivial embedding α of C_1 into $\hat{C}_1$. But for any elements $a, b \in C_1$, the type of b over a determines the type of b over $\alpha(a)$, so α is the identity, and the family of isomorphisms is closed under composition and inverse. Thus we may identify the Dedekind completions $\hat{C}$ and view Γ as a dense subset of $\hat{C}$ with a distinguished equivalence relation with dense classes.

Thus $\Gamma \cong S^{(n)}$.   □

LEMMA 20.3.20. *Let Γ be an imprimitive homogeneous 2-multi-tournament of shuffled type, with components of type $\mathbb{S}$. If there are at least three components, then Γ is properly shuffled.*

PROOF. Supposing the contrary, fix three classes C_1, C_2, C_3. Now the maps

$$f_{C_1,C_2} : \hat{C}_1 \to \hat{C}_2$$

considered in the previous argument are anti-automorphisms, and the composition

$$f_{C_3,C_1} \circ f_{C_2,C_3} \circ f_{C_1,C_2}$$

is an anti-automorphism of $\hat{C}_1$. In particular we have an anti-isomorphic embedding α of C_1 into $\hat{C}_1$ such that for $a, b \in C_1$, the type of b over a determines the type of b over $\alpha(a)$, and this is impossible.   □

LEMMA 20.3.21. *Let Γ be an imprimitive homogeneous 2-multi-tournament of shuffled type, with components of type $\mathbb{S}$. Then Γ is properly shuffled.*

PROOF. By Lemma 20.3.19 we may suppose that Γ has just two components S_1, S_2, and we suppose it is improperly shuffled. Note that by homogeneity there is an automorphism switching the two components and thus for either component C, when $a, b \notin C$ and $a \xrightarrow{1} b$ then $b' \cap C$ meets $a' \cap C$ in an initial segment.

In this case the map f_{S_2, S_1} from S_2 to the Dedekind completion $\bar{S}_1$ of S_1 is an anti-isomorphism. Thus we identify S_1, S_2 with dense subsets of the Dedekind completion $\bar{S}_1$ of S_1, with the arc relation on S_1 agreeing with that on $\bar{S}_1$, and the arc relation on S_2 the opposite of the arc relation on $\bar{S}_1$. Since we use the relation $\xrightarrow{2}$ to define this embedding, it agrees with the arc relation on $\bar{S}_1$.

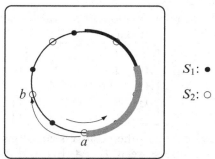

S_1: ●

S_2: ○

FIGURE 3. $\bar{S}_1$: $b' \cap S_1$ initial segment of $a' \cap S_1$.

If we enrich Γ by unary predicates naming S_1 and S_2 then this gives a unique structure, which is homogeneous.

However we will argue that in the language of 2-multi-tournaments this structure is not in fact homogeneous, and specifically that the two $\overset{1}{\sim}$-classes S_1, S_2 are invariant under the automorphism group.

For this, fix points $a \in S_1$ and $b \in S_2$, and consider the isomorphism type of the associated structures $\Gamma_a = (A_1, A_1^*, A_2, A_2^*)$ and $\Gamma_b = (B_2^*, B_2, B_1, B_1^*)$ where

$$A_1 = a^1; \qquad A_1^* = a^{1^{\mathrm{op}}}; \qquad A_2 = a^2; \qquad A_2^* = a^{2^{\mathrm{op}}}.$$
$$B_2^* = b^1; \qquad B_2 = b^{1^{\mathrm{op}}}; \qquad B_1 = b^2; \qquad B_1^* = b^{2^{\mathrm{op}}}.$$

We have chosen the notation so that in terms of the arc relation on $\bar{S}_1$ we have

$$A_1 = a' \cap S_1, \qquad A_2 = a' \cap S_2, \qquad A_1^* = {}'a \cap S_1, \qquad A_2^* = {}'a \cap S_2;$$
$$B_1 = b' \cap S_1, \qquad B_2 = b' \cap S_2, \qquad B_1^* = {}'b \cap S_1, \qquad B_2^* = {}'b \cap S_2,$$

which makes things somewhat easier to keep track of after embedding everything into $\bar{S}_1$.

Now an automorphism of Γ taking a to b would take Γ_a to Γ_b.

On the other hand Γ_a and Γ_b are variations on the 2-multi-tournament of properly shuffled type $\mathbb{Q}^{(4)}$ in which some of the arc relations have been reversed. Ini S_1, there are two a-definable ways to convert the tournament relation to a linear order with dense subsets a', $'a$: reverse arcs between the two parts, or on the two parts. As the arc relation on Γ_a is not induced directly by the arc relation on $\bar{S}_1$, we have two variants of this rule for converting Γ_1 into $\mathbb{Q}^{(4)}$, and two similar rules for Γ_b. We need to check that these are not the same two ways.

We may represent the relevant data by a matrix with entries ± 1, where $+1$ means the arc relation in question is not reversed, and -1 means it is reversed. We include the labels of the rows and columns of these matrices in our presentation of the matrix.

In matrix terms the re-orientations of Γ_a which result in a properly shuffled tournament may be represented as

$$\pm \begin{bmatrix} & A_1 & A_1^* & A_2 & A_2^* \\ A_1 & 1 & -1 & 1 & -1 \\ A_1^* & -1 & 1 & -1 & 1 \\ A_2 & 1 & -1 & -1 & -1 \\ A_2^* & -1 & 1 & -1 & -1 \end{bmatrix}.$$

The rule corresponding to the positive sign, as given by the matrix, is as follows. We reverse arcs between $A_1 \cup A_2$ (a') and $A_1^* \cup A_2^*$ ($'a$), and within A_2 or A_2^* (S_2).

In the case of Γ_b, if we labeled the columns and rows correspondingly with B_1, B_1^*, B_2, B_2^*, we would have the same pair of matrices, but as we have seen the order is different for Γ_b.

Thus in fact we get

$$\pm \begin{bmatrix} & B_2^* & B_2 & B_1 & B_1^* \\ B_2^* & -1 & -1 & -1 & 1 \\ B_2 & -1 & -1 & 1 & -1 \\ B_1 & 1 & -1 & 1 & -1 \\ B_1^* & -1 & 1 & -1 & 1 \end{bmatrix}.$$

Considering for example the first row in each case, we see that these are different. Thus a and b lie in different orbits of the automorphism group and we have a contradiction. $\square$

We may sum up as follows.

LEMMA 20.3.22. *Let Γ be an imprimitive homogeneous 2-multi-tournament of shuffled type, with n components of type $T = \mathbb{Q}$ or $\mathbb{S}$. Then up to a change of language Γ is isomorphic to $T^{(n)}$.*

20.3.2. General type, with a finite quotient. The hypotheses generally in use in the present subsection are the following.

(H1) Γ is an imprimitive homogeneous 2-multi-tournament with $n < \infty$ classes.

(H2) Γ is infinite, not a composition, and not of shuffled type.

Then Lemma 20.3.14 applies and Γ is of general type: if the components are of type $\mathbb{Q}$ or $\mathbb{S}$ then for each $\overset{1}{\sim}$-class C and each $u \notin C$, u^2 and $u^{2^{\mathrm{op}}}$ meet C in dense subsets.

In this case, our goal is to show that Γ is the generic de-symmetrization $n * T$ of the imprimitive directed graph $n \cdot T$. In other words, Γ is the generic imprimitive homogeneous 2-multi-tournament with n components, each of type T (Lemma 20.3.34).

We will derive this from an older classification result, the classification of homogeneous n-*partitioned tournaments* for n finite.

20.3.2.1. n-*Partitioned tournaments*.

DEFINITION 20.3.23. An n-*partitioned tournament* is a structure

$$\mathbb{T} = (T_1, \ldots, T_n)$$

with unary predicates P_i ($1 \le i \le n$), binary relations $\overset{i}{\to}$ ($1 \le i \le n$), and a finite number of additional binary relations $\overset{i,j}{\longrightarrow}_k$ ($1 \le i < j \le n$) such that the following conditions hold.

(a) P_i defines T_i.

(b) $(T_i, \overset{i}{\to})$ is a tournament (and all pairs related by $\overset{i}{\to}$ lie in T_i).

(c) The relations $\overset{i,j}{\longrightarrow}_k$ partition $T_i \times T_j$.

The classification of homogeneous n-partitioned tournaments is given in Cherlin [1998], where they are called n-*tournaments*; we will stick to our more explicit terminology here.

Our imprimitive 2-multi-tournaments with n classes (n finite) are similar to the n-partitioned tournaments with just two relations $\overset{i,j}{\longrightarrow}_k$ per choice of $i.j$, $1 \le i < j \le n$ corresponding to $\overset{2}{\to}$ and $\overset{2}{\leftarrow}$, and with all components isomorphic, after a non-trivial change of language: we expand the language by unary predicates for the individual $\overset{1}{\sim}$-classes. This substantially reduces the automorphism group, killing the permutations of the components, and splitting the relation $\overset{2}{\to}$ into many separate relations.

Our main task is to show that an imprimitive homogeneous 2-multi-tournament with a finite number of classes remains homogeneous when the $\overset{1}{\sim}$-classes are named, after which we will be able to invoke our classification of the homogeneous n-partitioned tournaments. It should be said that the classification of the homogeneous n-partitioned tournaments is more straightforward in the case that interests us, allowing only two "cross types" $\overset{i,j}{\longrightarrow}_k$ per pair (i, j), and—less importantly–requiring the components to be isomorphic. But the nature of the relevant classification result seems clearer when given in some generality. We will in fact recall the details in more generality than we need, but we will also set aside some issues irrelevant to our intended application.

The general classification for homogeneous n-partitioned tournaments involves a reduction from general finite n to the case $n = 2$. This goes as follows (we rephrase it in a more explicit form).

FACT 20.3.24 (Cherlin [1998, Proposition 9]). *For $n > 2$, a homogeneous n-partitioned tournament is determined by its transversals and its T-restrictions.*

Here the transversals are simply transversals to the equivalence relation with classes T_i, viewed as structures with the relevant relations $\xrightarrow{i,j}_k$ (here i, j will determined k). The 2-restrictions are the 2-partitioned tournaments with classes T_i, T_j and relations $\xrightarrow{i}$, $\xrightarrow{j}$, $\xrightarrow{i,j}_k$ for any choice of i, j.

Where things become more complex, or at least more detailed, is in the discussion of the homogeneous 2-partitioned tournaments. We now jump to the case which becomes relevant when we deal with homogeneous imprimitive 2-multi-tournaments of general type.

FACT 20.3.25 (Cherlin [1998, Propositions 4 & 5]). *Let $\mathbb{T}$ be a homogeneous 2-partitioned tournament with components T_1, T_2 of type $\mathbb{Q}$ or $\mathbb{S}$ and suppose that for $a \in T_1$ and any cross-type p, a^p is dense in T_2. Then $\mathbb{T} \cong \Gamma_k(T_1, T_2)$ is the generic 2-partitioned tournament with k cross types, for some k.*

Here the components are not required to be isomorphic, but that is the only case that will concern us.

For homogeneous 2-partitioned tournaments with components of type T^∞ things are somewhat more complicated. First of all, we set aside *diagonal type*, in which one of the relations $\xrightarrow{1,2}_k$ gives a bijection between the components. The relevant classification runs as follows—and this is where the extra generality of the setting will lead us into a technical digression. We give the statement and then explain the notation used.

FACT 20.3.26 (Cherlin [1998, Proposition 7]). *Let $\mathbb{T} = (T_1, T_2)$ be a homogeneous 2-partitioned tournament, not of diagonal type, with $T_1, T_2 \cong T^\infty$, and let P be the set of cross types in the language of $\mathbb{T}$. Then there is a partition*

$$\mathcal{P} = (P^0, P^+, P^-)$$

of P such that

$$\mathbb{T} \cong \Gamma(\mathcal{P})$$

up to a change of language.

We use the term "partition" loosely here: we allow some of the specified sets to be empty.

The notation $\Gamma(P)$ refers to the following construction—but we should say in advance that when $|P| = 2$ our partition $\mathcal{P}$ will be $(P, \emptyset, \emptyset)$ and $\Gamma(\mathcal{P})$ will be the generic 2-partitioned tournament with cross types in P, so this all falls away. The following definition is the case of Cherlin [1998, Definition 3] which is of interest here.

Definition 20.3.27. Let P be a set of $k \geq 1$ cross types (binary relation symbols). Let $\mathcal{P} = (P^0, P^+, P^-)$ be a partition of P into three sets, where P^0 is non-empty but P^+, P^- may be empty. Then

$$\mathcal{A}(\mathcal{P})$$

denotes the class of finite 2-partitioned tournaments $A = (A_1, A_2)$ satisfying the following conditions.

1. Cross types come from P (every pair in $A_1 \times A_2$ satisfies a unique relation in P).
2. There are no substructures of A of the form $(a_1, b_2^- b_2^+)$ or $(b_1^- b_1^+, a_2)$ with $b_i^- \to b_i^+$ and

$$\text{tp}(a_1 b_2^\pm) \in P^\pm \text{ or}$$
$$\text{tp}(b_1^\pm a_2) \in P^\pm.$$

In words: over a parameter a in T_1 or T_2, the points in the other component related positively to a dominate those related negatively to a.

With P^0 non-empty the class $\mathcal{A}(\mathcal{P})$ is an amalgamation class and $\Gamma(\mathcal{P})$ will denotes its Fraïssé limit. $\Gamma(\mathcal{P})$ is simply the generic 2-partitioned tournament with $k = |P|$ cross types unless P^+ and P^- are non-empty, and since we also require P^0 non-empty, this case first occurs for $k = 3$. So it will not appear in our application here.

We give that application now.

Lemma 20.3.28. *Let Γ be an imprimitive homogeneous 2-multi-tournament with $n < \infty$ components, of general type. Suppose the following conditions are satisfied.*

1. *If Γ^+ is the expansion of Γ by names for the components of Γ, then Γ^+ is homogeneous.*
2. *Every tournament of order n embeds as a transversal into Γ.*
3. *$\text{Aut}(\Gamma)$ induces the full symmetric group on the n components of Γ*

Then Γ is the generic imprimitive homogeneous 2-multi-tournament with n components of type T_1.

Proof. With a slight change of language Γ^+ may be viewed as an n-partitioned tournament, replacing the relations $\xrightarrow{1}$ and $\xrightarrow{2}$ by the various relations induced on each pair of components (not necessarily distinct).

By our second and third hypotheses, any tournament

$$A = (a_1, \ldots, a_n)$$

of order n embeds as a transversal into Γ with a_i mapping to the i-th component. In other words, every possible transversal occurs.

By Fact 20.3.24 it suffices to check that the 2-restrictions of Γ^+ are generic imprimitive with two components of type T_1.

When the components are of type $\mathbb{Q}$ or $\mathbb{S}$ then Fact 20.3.25 applies directly. So suppose the components are of type T^∞. We refer to Fact 20.3.26.

In that result, the term *diagonal type* refers to the case in which one of the cross relations defines a bijection β between the two components. This cannot happen when we have just two cross types, as we would have the three relations $y = \beta(x)$, $y \xrightarrow{1} \beta(x)$, $\beta(x) \xrightarrow{1} y$.

Furthermore, as there are only two cross types, the decomposition

$$\mathcal{P} = P^0 \sqcup P^+ \sqcup P^-$$

with P^0 non-empty must have P^+ or P^- empty and thus $\Gamma(\mathcal{P})$ is the generic imprimitive homogeneous 2-partitioned tournament with two classes.

The result follows. $\square$

20.3.2.2. *Structure of u^2.*

LEMMA 20.3.29. *Let Γ be an imprimitive homogeneous 2-multi-tournament with $n < \infty$ components, of general type. Let C_1, C_2 be two $\overset{1}{\sim}$-classes and let $A \subseteq C_1$ be finite and transitive (a linear order). Let*

$$A' = \{x \in \Gamma \mid x \in a' \ (all \ a \in A)\}.$$

Then $A' \cap C_2$ is infinite.

PROOF. Supposing the contrary, take a minimal counterexample A. For any $m < \infty$ set

$$F_m(A) = \{C \in \Gamma/\overset{1}{\sim} \mid C \neq C_1, |A' \cap C| < m\}.$$

Choose m so that $F_m(A)$ is nonempty.

CLAIM 1. *For $A_1, A_2 \cong A$ in C_1, $F_m(A_1) = F_m(A_2)$.*

Take $B_i \cong A$ in C_1 for $i \in \mathbb{N}$ such that $B_i \xrightarrow{1} B_j$ for $i < j$. There is then $i < j$ for which $F_m(B_i) = F_m(B_j)$. If $A_1 \xrightarrow{1} A_2$ then by homogeneity we conclude $F_m(A_1) = F_m(A_2)$. In general, we take $A_3 \cong A$ with $A_1, A_2 \xrightarrow{1} A_3$ and apply the previous case. This proves the claim.

Thus we may write $F_m(C_1) = F_m(A)$ for any choice of A within C_1. Then Lemma 20.3.8 implies that $F_m(C_1)$ consists of all classes in $\Gamma/\overset{1}{\sim}$ other than C_1.

Now let $A = A_0 \cup \{a^*\}$, where $A_0 \xrightarrow{1} a^*$. Fix a $\overset{1}{\sim}$-class $C_2 \neq C_1$. Then $A_0' \cap C_2$ is infinite.

Take a subset B of $\{b \in C_1 \mid A_0 \to b\}$ with $|B| = m$. For $b \in B$, by hypothesis $b' \cap (A_0' \cap C_2)$ is finite, so $'B \cap (A_0' \cap C_2)$ is infinite. Take $A^* \subseteq {}'B \cap (A_0' \cap C_2)$ isomorphic to A. Then $B \subseteq (A^*)'$, $|B| = m$, but $C_1 \in F_m(A^*)$, a contradiction. $\square$

In the next lemma we use the fact that if the generic tournament T^∞ is partitioned into finitely many pieces then every finite tournament embeds into one of the pieces—actually it is easy to show that T^∞ itself embeds into one of the pieces; this stronger property is called *indivisibility*. Cf. El-Zahar and Sauer [1991], [1993].

LEMMA 20.3.30. *Let Γ be an imprimitive homogeneous 2-multi-tournament of general type. Then for $u \in \Gamma$ the components of u^2 are isomorphic to the components of Γ.*

PROOF. If the components of Γ are of type $\mathbb{Q}$ or $\mathbb{S}$, then by assumption the components of u^2 are dense in the components of Γ, and the result follows. So we may suppose the components of Γ are of type T^∞.

If the result fails, then there is a finite substructure A of T^∞ which does not embed in u^2 for $u \in \Gamma$. Fix two $\overset{1}{\sim}$-classes C_1, C_2 and take A embedded into C_2. Then $'A \cap C_1$ is empty, that is, C_1 is the union of the sets $a' \cap C_1$ as a varies over A. As this is a finite union, one of the sets $a' \cap C_1$ contains a copy A_1 of A, and this is a contradiction. □

LEMMA 20.3.31. *Let Γ be an imprimitive homogeneous 2-multi-tournament of general type. Suppose that $|\Gamma/\overset{1}{\sim}| = n$ with $3 \le n < \infty$. Then for $u \in \Gamma$, the homogeneous 2-multi-tournament u^2 is not a composition.*

PROOF. If the lemma fails, fix a $\overset{1}{\sim}$-class C and for $x \in C$ let T_x be the tournament induced on $x^2/\overset{1}{\sim}$ by $\overset{2}{\longrightarrow}$.

By Lemma 20.3.8, x^2 meets every $\overset{1}{\sim}$-class other than C, so this gives a corresponding (and isomorphic) tournament structure on $(\Gamma \setminus C)/\overset{1}{\sim}$. Let T_x^* denote the latter.

For any two elements $x, y \in C$, $x^2 \cap y^2$ also meets every $\overset{1}{\sim}$-class other than C, by Lemma 20.3.29. It follows that $T_x^* = T_y^*$, and this tournament may be denoted T_C.

In particular for three distinct $\overset{1}{\sim}$-classes C, C_1, C_2, no automorphism of Γ takes (C, C_1, C_2) to (C, C_2, C_1). With $x \in C$ fixed, take

$$A_1 = x^2 \cap C_1; \qquad\qquad A_2 = x^2 \cap C_2;$$
$$B_1 = x^{2^{\mathrm{op}}} \cap C_1; \qquad\qquad B_2 = x^{2^{\mathrm{op}}} \cap C_2.$$

If there are arcs (a_1, b_2) and (a_2, b_1) having the same orientation with $a_i \in A_i$, $b_i \in B_i$, then there is an automorphism taking (x, a_1, b_2) to (x, a_2, b_1), which then takes (C, C_1, C_2) to (C, C_2, C_1), for a contradiction. Thus the arcs between A_1 and B_2 have a fixed orientation, which agrees with the orientation of arcs between B_1 and A_2. On the other hand, the arcs between A_1 and A_2 also have a fixed orientation, and as Γ is not a composition, these orientations are opposite.

Furthermore, if there are arcs of two orientations between B_1 and B_2, then again we can find an automorphism carrying C to itself and switching C_1, C_2.

We conclude that there are at just two types realized in C_1 over C_2, namely those realized by points of A_1, and those realized by points of B_1. As $\Gamma/\overset{1}{\sim}$ is finite, there are in fact only finitely many types realized in C_1 over the rest of

Γ. But C_1 is primitive, and so all points of C_1 realize a single type over the rest of Γ, and Γ is itself a composition, which is a contradiction.   □

LEMMA 20.3.32. *Let Γ be an imprimitive homogeneous 2-multi-tournament of general type, with components of type $\mathbb{Q}$ or $\mathbb{S}$. Suppose that $|\Gamma/\overset{1}{\sim}| = n$ with $n < \infty$. Then for any two $\overset{1}{\sim}$-classes C_1, C_2, and any finite transitive substructure L of C_1, every 1-type over L is realized in C_2 (that is, every subset of L is of the form $a' \cap L$ with $a \in C_2$).*

PROOF. Take $u \in C_2$ and use the density of $c' \cap C_1$ and $'c \cap C_1$ in C_1 to find a transitive substructure L_1 of C_1 so that uL_1 has the desired isomorphism type.

Then take vertices $c_1 \in L_1' \cap C_2$ and $c \in L' \cap C_2$, and take an automorphism carrying $c_1 L_1$ to cL. Then u goes to a realization of the desired type in C_2.   □

LEMMA 20.3.33. *Let Γ be an imprimitive homogeneous 2-multi-tournament of general type. Suppose that $|\Gamma/\overset{1}{\sim}| = n$ with $n < \infty$. Then for $u \in \Gamma$, u^2 is of general type, with components as in Γ.*

PROOF. We dealt with the components in Lemma 20.3.30, and by Lemma 20.3.31, u^2 is not a composition.

So it suffices to show that u^2 is not shuffled. In particular, we may suppose the components of Γ are of type $\mathbb{Q}$ or $\mathbb{S}$.

We perform an amalgamation. Define

$$L = \{u_1, u_2, u_3, u_4\}, \quad L \text{ transitive with } u_1 \xrightarrow{1} u_2 \xrightarrow{1} u_3 \xrightarrow{1} u_4;$$

$$A_0 = Lu, \quad u \xrightarrow{2} u_1, u_2, u_4; u \xleftarrow{2} u_3;$$

$$A_1 = A_0 \cup \{a_1\}, \quad a_1 \xrightarrow{2} u_1, u_2, u_4, a_1 \xleftarrow{2} u_3, a_1 \xrightarrow{2} u \text{ or } u \xrightarrow{2} a_1;$$

$$A_2 = A_0 \cup \{a_2\}, \quad a_2 \xrightarrow{2} u_1, u_3, u_4, a_2 \xleftarrow{2} u_2, a_2 \xrightarrow{1} u \text{ or } u \xrightarrow{1} a_2.$$

Note that the $\overset{1}{\sim}$-classes involved here are L, a_1, and $a_2 u$.

If A_1, A_2 embed into Γ then we amalgamate them over A_0. Then $a_1 \xrightarrow{2} a_2$ or $a_2 \xrightarrow{2} a_1$, and correspondingly

$$a_1 \to (a_2, u_1, u_2, u_4) \text{ or } a_2 \to (a_1, u_1, u_3, u_4)$$

which will show that a_1' or a_2' is not convex.

So it remains to check that A_1 and A_2 embed into Γ.

Take three $\overset{1}{\sim}$-classes C, C_1, C_2 and $L \subseteq C$. It suffices to realize the type of a_1 over L in C_1, and to realize the types of a_2 and u over L in C_2, using Lemma 20.3.32.   □

20.3.2.3. *Identification: General type (finite quotient).*

LEMMA 20.3.34. *Let Γ be an imprimitive homogeneous 2-multi-tournament of general type, with components of type T ($\mathbb{Q}$, $\mathbb{S}$, or T^∞). Suppose that $|\Gamma/\overset{1}{\sim}| = n$*

with $n < \infty$. Then Γ is the generic homogeneous 2-multi-tournament with n components of type T.

PROOF. We will argue by induction on n.

For $u \in \Gamma$, u^2 is of general type, with $(n-1)$ components of type T, by Lemma 20.3.33. So we suppose that u^2 is the generic imprimitive homogeneous 2-multi-tournament with $(n-1)$ components of type T.

Let Γ^+ be the expansion of Γ by names for the components.

CLAIM 1. Γ^+ *is homogeneous.*

We suppose $\alpha : A \xrightarrow{\cong} B$ with A, B finite, by an isomorphism respecting the labels on the $\overset{1}{\sim}$-classes. Then A and B meet the same $\overset{1}{\sim}$-classes of Γ. We must find an automorphism of Γ leaving the components invariant and taking A to B.

We argue by induction on the number of $\overset{1}{\sim}$-classes disjoint from A in Γ. If A meets all $\overset{1}{\sim}$-classes of Γ then any extension of α to an automorphism of Γ will be an automorphism of Γ^+.

Now suppose A does not meet the $\overset{1}{\sim}$-class C of Γ. Take $u \in C$. There is a copy of $A \cup B$ in u^2 and hence there is some $v \in \Gamma$ with $A \cup B \subseteq v^2$. Replacing A, B by $A_1 = A \cup \{v\}$, $B_1 = B \cup \{v\}$, we reduce the number of $\overset{1}{\sim}$-classes disjoint from A_1 and conclude by induction.

CLAIM 2. *Every $\xrightarrow{2}$-tournament K of order n embeds as a transversal into Γ.*

Let $a, b \in K$, $K_0 = K \setminus \{a, b\}$. Construct $A_0 = K_1 \cup K_2 \cup \{u\}$ so that

$$K_1, K_2 \cong K_0, \qquad\qquad u \xrightarrow{2} K_1 K_2;$$

$$k_1 \xrightarrow{1} k_2 \text{ if the elements } k_1 \in K_1 \text{ and } k_2 \in K_2$$
$$\text{correspond to the same element of } K_0.$$

Adjoin a_1, a_2 with

$$u \xrightarrow{1} a_1, \qquad\qquad\qquad\qquad u \xrightarrow{2} a_2;$$

$$\text{tp}(a_1/K_1) = \text{tp}(a/K_0), \qquad \text{tp}(a_2/K_1) = \text{tp}(b/K_0);$$

$$\text{tp}(a_1/K_2) = \text{tp}(b/K_0), \qquad \text{tp}(a_2/K_2) = \text{tp}(a/K_0).$$

Amalgamating $A_1 = A_0 \cup \{a_1\}$ with $A_2 = A_0 \cup \{a_2\}$ over A_0 gives $a_1 \xrightarrow{2} a_2$ or the reverse, with correspondingly

$$K_1 a_1 a_2 \text{ or } K_2 a_2 a_1 \cong K.$$

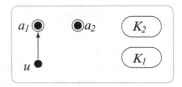

We claim that the factors A_1, A_2 embed into Γ. Note however that at this point we cannot control which $\overset{1}{\sim}$-classes of Γ the elements corresponding to a, b lie in.

The factor A_1 meets only $(n-1)$ $\overset{1}{\sim}$-classes of Γ, so it embeds into Γ.

The factor A_2 is of the form $u \overset{2}{\longrightarrow} (a_2 K_1 K_2)$ and $a_2 K_1 K_2$ meets only $(n-1)$ $\overset{1}{\sim}$-classes of Γ. Hence A_2 also embeds into Γ. Therefore we can make the indicated amalgamation to embed K into Γ.

The claim follows.

CLAIM 3. Aut Γ *induces the full symmetric group on* $\Gamma/\overset{1}{\sim}$.

Let $L = (a_1, \ldots, a_n)$ be a transitive tournament and $T_1, \ldots, T_n$ an enumeration of the components. Then we can embed L into Γ with a_i mapping to T_i, by sending a_1 to $u \in T_1$ and then applying induction to u^2. Call the image L_1.

Similarly we can map L into $L_\sigma \subseteq \Gamma$ so that a_i goes into T_{i^σ} for any permutation σ. The isomorphism $L_1 \cong L_\sigma$ extends to an automorphism of Γ which acts on the components like σ.

This proves the claim.

Now apply Lemma 20.3.28 to conclude. $\qquad\qquad\qquad\qquad\qquad\qquad\square$

20.3.3. Infinite quotients and the parity constraint. It remains to classify the homogeneous 2-multi-tournaments with infinite quotient. This case subdivides, as we have the semi-generic 2-multi-tournaments with specified classes. So we adopt the following terminology for the sake of brevity.

DEFINITION 20.3.35. Let Γ be a homogeneous 2-multi-tournament.

1. Γ is of *semi-generic type* if Γ is of general type, $\Gamma/\overset{1}{\sim}$ is infinite, and the parity constraint is satisfied.

2. Γ is of *generic type* if Γ is of general type, $\Gamma/\overset{1}{\sim}$ is infinite, and the parity constraint is not satisfied.

LEMMA 20.3.36. *Let* Γ *be an imprimitive homogeneous 2-multi-tournament of semi-generic type. Then for* $u \in \Gamma$, u^2 *is not a composition.*

PROOF. Let C be the $\overset{1}{\sim}$-class of u. Suppose u^2 is a composition.

Take C_1, C_2 two more $\overset{1}{\sim}$-classes with $u^2 \cap C_1 \overset{2}{\longrightarrow} u^2 \cap C_2$. Set

$$A_1 = u^2 \cap C_1, \qquad A_2 = u^{2^{\mathrm{op}}} \cap C_1, \qquad B_1 = u^2 \cap C_2, \qquad B_2 = u^{2^{\mathrm{op}}} \cap C_2.$$

If there are $a_1 \in A_1$, $a_2 \in A_2$, $b_1 \in B_1$, $b_2 \in B_2$ such that (a_1, b_2) and (b_1, a_2) have the same orientation, then there is an automorphism fixing u and interchanging C_1, C_2, which gives a contradiction.

It follows that $A_1 \times B_2$ and $A_2 \times B_1$ each realize one 2-type, and these types are the same.

By the parity constraint we also have $A_2 \overset{2}{-} B_2$ and thus each of C_1, C_2 realizes two types over the other. Thus C_1 realizes only two types over $\Gamma \setminus (C \cup C_1)$, and varying C shows that C_1 realizes only two types over $\Gamma \setminus C_1$. But C_1 is primitive, so this is a contradiction.

Take $a_1, a_2 \in A_1$ with $a_1 \overset{1}{\to} a_2$, and $b_1, b_2 \in C_1$ with $b_1 \overset{1}{\to} b_2$. Using the parity constraint it follows easily that there are $c_1, c_2 \in C_2$ with $a_1, a_2 \overset{2}{\to} c_1$, and $b_1, b_2 \overset{2}{\to} c_2$. There is an automorphism taking (a_1, a_2, c_1) to (b_1, b_2, c_2). As (a_1, a_2) realize the same type over C_2, the same applies to (b_1, b_2). Thus all $a \in C_1$ realize the same type over C_2, Similarly all $a \in C_2$ realize the same type over C_1. But then Γ is a composition. □

LEMMA 20.3.37. *Let Γ be an imprimitive homogeneous 2-multi-tournament of semi-generic type. Then Γ is the semi-generic tournament with components of the given type.*

PROOF. We will prove the following by induction on n.

If A is a finite imprimitive 2-multi-tournament with $n \overset{1}{\sim}$-classes, and A satisfies the parity constraint, then A embeds into Γ. $(*_n)$

So we fix A with $n \overset{1}{\sim}$-classes and we assume the claim holds when fewer classes are involved.

CLAIM 1. *Without loss of generality, A has at most one nontrivial $\overset{1}{\sim}$-class.*

We take a transversal K to A and view A as the amalgam of structures $A_C = K \cup C$, with C varying over $\overset{1}{\sim}$-classes in A. By the parity constraint, A is the unique amalgam of these structures.

Now suppose the following.

$(\star)$ $n = 2$.

If the components of Γ are of type $\mathbb{Q}$ or $\mathbb{S}$ then as $u^2 \cap C$ and $u^{2^{op}} \cap C$ are dense in C for $u \notin C$, the claim follows. So suppose the components of Γ are of type T^∞.

We consider $\Gamma^* = (u^2 \cap C, u^{2^{op}} \cap C)$ for $u \notin C$. This is a homogeneous 2-partitioned tournament with labeled parts, whose components are of type T^∞, and with at most two cross types. It suffices to show that Γ^* is generic with two cross types.

If there is just one cross type then $u^2 \cap C \overset{1}{\to} u^{2^{op}} \cap C$ or the reverse. But $C \cong T^\infty$ has no such decomposition with nonempty parts. So there are two cross types. Now apply Fact 20.3.26.

This disposes of the case $n = 2$, and the result to be proved is trivial for $n = 1$. So we now suppose the following.

($\star$) $n \geq 3$.

We have $A = KC$ with K a $\xrightarrow{2}$-tournament and C a single $\xrightarrow{1}$-class. Fix $a, b \in K$ and let $K_0 = K \setminus \{a, b\}$. Take $K_1 C_1, K_2 C_2 \cong K_0 C$ and form $A_0 = K_1 K_2 uv$ with

$$uv C_1 C_2 \text{ a } \overset{1}{\sim}\text{-class,}$$

$$k_1 \xrightarrow{1} k_2 \text{ when } k_1, k_2 \text{ correspond to the same element of } K_0,$$

and so that A_0 satisfies the parity constraint.

Now amalgamate $A_1 = A_0 \cup \{a_1\}$ with $A_2 = A_0 \cup \{a_2\}$ over A_0, where

$$a_1 K_1 C_1 \cong a_2 K_2 C_2 \cong a K_0 C; \qquad a_2 K_1 C_1 \cong a_1 K_2 C_2 \cong b K_0 C;$$

$$a_1 \xrightarrow{2} u, v, \qquad\qquad v \xrightarrow{2} a_2 \xrightarrow{2} u.$$

Now A_1, A_2 involve $(n - 1)$ $\overset{1}{\sim}$-classes and satisfy the parity constraint, hence embed into Γ.

In their amalgam, the points u, v force a_1, a_2 to lie in distinct $\overset{1}{\sim}$-classes. So $a_1 \xrightarrow{2} a_2$ or $a_2 \xrightarrow{2} a_1$, and correspondingly $(a_1 a_2 K_1 C_1)$ or $(a_2 a_1 K_2 C_2)$ is isomorphic to $KC = A$. $\qquad\square$

20.3.4. Generic type: 2-restrictions. Now we consider the case in which Γ is of generic type: that is, of general type, with an infinite quotient, and not satisfying the parity constraint. We begin by showing that the restriction of Γ to two components is a generic imprimitive 2-multi-tournament with two classes (Lemma 20.3.41). But this requires some preliminaries.

By Lemma 20.3.30 the components of u^2 are as in Γ, for $u \in \Gamma$.

LEMMA 20.3.38. *Let Γ be an imprimitive homogeneous 2-multi-tournament of generic type with components of type $\mathbb{Q}$. Then for any two $\overset{1}{\sim}$-classes C_1, C_2 and any finite configuration (L_1, L_2) with two transitive $\overset{1}{\sim}$-classes, there is an embedding of $L_1 L_2$ into Γ which takes L_1 into C_1 and L_2 into C_2.*

PROOF. By amalgamations with unique solutions we reduce to the case $|L_1| = |L_2| = 2$ and we adopt the notation

$$L_1 = \{a_1, a_2\}, \qquad\qquad L_2 = \{b_1, b_2\};$$

$$a_1 \xrightarrow{1} a_2, \qquad\qquad b_1 \xrightarrow{1} b_2.$$

Inserting u with $a_1 \xrightarrow{1} u \xrightarrow{1} a_2$ and $u \xrightarrow{2} b_1, b_2$, and determining the type of (a_1, a_2) by an amalgamation, we reduce to the cases in which $a_1 \xrightarrow{2} b_1, b_2$ or $a_2 \xrightarrow{2} b_1, b_2$.

Suppose for example $a_1 \xrightarrow{2} b_1, b_2$. Consider $(a_1' \cap C_1, a_1' \cap C_2)$, a homogeneous 2-partitioned tournament with labeled parts having components of type $\mathbb{Q}$ and two cross types. It follows that if a_2 realizes the same type over b_1 and b_2, this configuration is realized; similarly if we begin with $a_2 \xrightarrow{2} b_1, b_2$. Accordingly only four configurations come into consideration.

$(I) \quad a_1 \xrightarrow{2} b_1, b_2: \quad (IA) \quad b_2 \xrightarrow{2} a_2 \xrightarrow{2} b_1; \quad (IB) \quad b_1 \xrightarrow{2} a_2 \xrightarrow{2} b_2;$

$(II) \quad a_2 \xrightarrow{2} b_1, b_2: \quad (IIA) \quad b_1 \xrightarrow{2} a_1 \xrightarrow{2} b_2; \quad (IIB) \quad b_2 \xrightarrow{2} a_1 \xrightarrow{2} b_1.$

Reversal of $\xrightarrow{1}$ interchanges (IA) and (IIA), as well as (IB) and (IIB).

We now consider the two amalgamation diagrams shown.

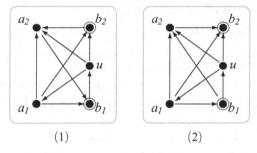

(1)   (2)

These force $b_1 \to b_2$ and hence produce (IA) or (IB) respectively.

Consider their factors.

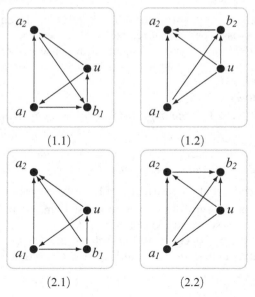

(1.1)   (1.2)

(2.1)   (2.2)

Suppose configuration (IB) embeds into Γ. The configuration (1.2) is isomorphic to (IB). Viewing the configuration (1.1) as an amalgamation diagram with the type of (a_1, u) left to be determined produces either the configuration (IA) or the factor (1.1). As the configurations $(1.1, 1.2)$ give (IA) by amalgamation, it follows that

$$(IB) \Longrightarrow (IA);$$

that is, if (IB) embeds in Γ, then (IA) does as well.

Similarly, (2.2) can be viewed as an amalgamation diagram to determine the type of (a_1, u), and (2.1) is isomorphic to (IIA). So we find

$$(IIA) \Longrightarrow (IB).$$

Reversing $\xrightarrow{1}$, these relations become

$$(IIB) \Longrightarrow (IIA); \qquad\qquad (IA) \Longrightarrow (IIB).$$

Now since Γ does not satisfy the parity constraint, one of the configurations (IA, IB, IIA, IIB) embeds into Γ, and it follows that they all do. $\qquad\square$

Lemma 20.3.39. *Let Γ be an imprimitive homogeneous 2-multi-tournament of generic type with components of type $\mathbb{S}$. Then for any two $\overset{1}{\sim}$-classes C_1 and any finite configuration (L_1, L_2) with two transitive $\overset{1}{\sim}$-classes, there is an embedding of $L_1 L_2$ into Γ which takes L_1 into C_1 and L_2 into C_2.*

Proof. Fix $c_1 \in C_1$ and $c_2 \in C_2$. Let $\Gamma_{c_1, c_2} = (U_1, U_2)$ with

$$U_1 = c_1^1 \cap c_2^{2^{op}}; \qquad\qquad U_2 = c_2^1 \cap c_1^{2^{op}}.$$

This is a homogeneous 2-partitioned tournament with labeled components of type $\mathbb{Q}$, and two cross types, and it realizes both cross types.

In fact, for $u \in U_2$, as u^2 and $u^{2^{op}}$ meet C_1 in dense subsets, they meet U_1 in dense subsets. So Γ_{c_1, c_2} is a 2-partitioned tournament of generic type with components of type $\mathbb{Q}$.

The result follows. $\qquad\square$

Lemma 20.3.40. *Let Γ be an imprimitive homogeneous 2-multi-tournament of generic type. Then for any two $\overset{1}{\sim}$-classes C_1, C_2 and any finite configuration (A, B) with two $\overset{1}{\sim}$-classes, each embedding in the components of Γ, there is an embedding of AB into Γ which takes A into C_1 and B into C_2.*

We do not yet claim that the restriction of Γ to two $\overset{1}{\sim}$-classes is homogeneous.

Proof. Let T be the type of the components. Lemma 20.3.38 handles the case in which $T = \mathbb{Q}$, so we suppose T is $\mathbb{S}$ or T^∞.

By hypothesis the parity constraint is violated in $C_1 \cup C_2$. Up to a change of language we may take $a_1, b_1 \in C_1$ and $a_2, b_2 \in C_2$ so that

$$a_1 \xrightarrow{2} a_2, b_2; \qquad b_2 \xrightarrow{2} b_1 \xrightarrow{2} a_2; \qquad a_1 \xrightarrow{1} b_1.$$

Let $p = \text{tp}(b_2/a_2)$, and set

$$\Gamma^* = (U, V) \text{ with } U = (a_1^1 \cap b_1^{2^{\text{op}}}, V = a_1^2 \cap a_2^p).$$

CLAIM 1. Γ^* *has components of type T if $T = T^\infty$, and type $\mathbb{Q}$ if $T = \mathbb{S}$.*

We have $b_1^{2^{\text{op}}} \cap C_1, a_1^2 \cap C_2 \cong T$, and we compute a_1^1 or a_2^p there.

CLAIM 2. Γ^* *realizes both cross types $\xrightarrow{2}$ and $\xleftarrow{2}$ from U to V.*

The type $b_2 \xrightarrow{2} b_1$ is realized by hypothesis.

The type $b_1 \xrightarrow{2} b_2$ corresponds to $a_1, a_2 \xrightarrow{2} b_1, b_2$.

If there is $a \in a_1^1 \cap C_1$ with $a^2 \cap a_1^2 \cap C_2$ containing two points, this suffices. Assume the contrary. Fix $c_1, c_2 \in a_1^1 \cap C_1$. Then with at most two exceptions, for $d \in a_1^2 \cap C_2$ we have $d \xrightarrow{2} c_1, c_2$. Take $d_1, d_2 \in a_1^2 \cap C_2$ with $d_1, d_2 \xrightarrow{2} c_1, c_2$. Take an automorphism carrying d_1, c_1 to a_1, d_1. Then a_2, d_2 go to some u, v with

$$a_1 \xrightarrow{1} v; \qquad d_1 \xrightarrow{1} u; \qquad a_1, v \xrightarrow{2} d_1, u.$$

This proves the claim.

CLAIM 3. Γ^* *is a generic 2-partitioned (labeled) tournament.*

Since Γ^* realizes both cross types, the only alternative, according to Cherlin [1998], would be that the components of Γ^* are local orders and they are shuffled (cf. Facts 20.3.25 and 20.3.26 and the discussion in §20.3.2.1).

But by Lemma 20.3.39 every finite configuration (L_1, L_2) with L_1, L_2 transitive embeds into Γ. This gives a contradiction and proves the claim.

Now if $T = T^\infty$ then as Γ^* has the same components as Γ, our lemma is proved. So it remains to complete the analysis for the case

$$T = \mathbb{S}.$$

CLAIM 4. *Let (A, L) be finite with A, L $\xrightarrow{1}$-classes, A a local order and L transitive. Then (A, L) embeds into Γ.*

Define $\mathcal{A}^*$ as the class

$\{A \mid A$ is a finite $\xrightarrow{1}$-local order and any finite extension (A, L)

with L a transitive $\xrightarrow{1}$-tournament and cross types $\xrightarrow{2}, \xleftarrow{2}$

embeds into $\Gamma.\}$

Then $\mathcal{A}^*$ is an amalgamation class containing all finite transitive $\xrightarrow{1}$-tournaments. Hence our claim reduces to the claim that

$$C_3 \in \mathcal{A}^*.$$

So we consider (C_3, L) with $C_3 = (a, b, c)$ and L finite and transitive. We make an amalgamation to determine the type of (a_1, a_2) in a configuration

$a_1 a_2 u v L_1 L_2$ in which

$$a_1 \xrightarrow{1} v \xrightarrow{1} a_2 \xrightarrow{1} u \xrightarrow{1} a_1; \qquad u \xrightarrow{1} v; \qquad\qquad L_1 \xrightarrow{1} L_2;$$
$$aL \cong a_1 L_1 \cong a_2 L_2; \qquad\qquad bL \cong a_2 L_1 \cong a_1 L_2; \quad cL \cong u L_1 \cong u L_2.$$

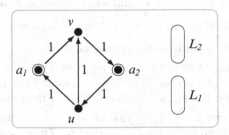

Thus in the amalgam $a_1 \xrightarrow{1} a_2$ or $a_2 \xrightarrow{1} a_1$ and correspondingly $(a_1 a_2 u L_1)$ or $(a_2 a_1 v L_2)$ will be isomorphic to $C_3 L$. So it suffices to check that the factors omitting a_2 or a_1 embed into Γ, in some matching forms (we have not yet settled the types of v over L_1 or u over L_2).

The factor omitting a_2 is the union of two transitive $\overset{1}{\sim}$-classes and embeds in Γ^*, hence in Γ. So it suffices to find a suitable form for the factor omitting a_1.

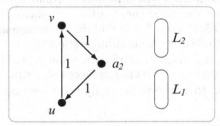

We may obtain a suitable factor as an amalgam in which the type of v over L_1 remains to be determined. The factor omitting v is then a union of two linear classes, so this leaves us with the factor omitting L_1.

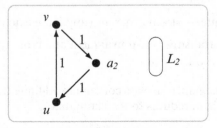

We view the diagram as an amalgamation in which the type of u over L_2 remains to be determined. Then both factors embed in Γ and we may conclude. This proves the claim.

CLAIM 5. *Let (A, B) be finite with A, B $\xrightarrow{1}$-classes, and each a local order. Then (A, B) embeds into Γ.*

We consider the class $\mathcal{A}^{**}$ of finite A obeying the stated conditions for general B. This is then an amalgamation class, and by the previous claim it contains all finite transitive $\xrightarrow{1}$-tournaments. It suffices to show that it contains C_3. So we know that any extension (A, L) with $A \cong C_3$ and L transitive embeds into Γ and we need to deal with (A, B) where B is a finite local order.

We repeat the amalgamation argument used in the previous claim, replacing L_1, L_2 by B_1, B_2 and dropping the condition $L_1 \xrightarrow{1} L_2$ in favor of: $B_1 \cup B_2$ is a local order. This goes as before.

The last claim completes the proof of the lemma in the last remaining case, namely when $T = \mathbb{S}$. □

LEMMA 20.3.41. *Let Γ be an imprimitive homogeneous 2-multi-tournament of generic type and components of type T. Then for any two $\overset{1}{\sim}$-classes C_1, C_2 the restriction of Γ to $C_1 \cup C_2$ is a generic 2-multi-tournament with two components of type T.*

PROOF. By the previous lemma, it suffices to show that the restriction of Γ to $C_1 \cup C_2$ is homogeneous.

Let $A, B \subseteq C_1 \cup C_2$ be finite and $f : A \xrightarrow{\cong} B$. If A meets C_1 and C_2 then any automorphism of Γ extending f takes $C_1 \cup C_2$ to itself.

There remains the case in which $A \subseteq C_1$ and B is contained in C_1 or C_2. Since we may interchange C_1, C_2 by an automorphism of Γ, we may suppose $B \subseteq C_1$.

Let L be finite and transitive so that (A, L) realizes every 1-type involving $\xrightarrow{2}, \xleftarrow{2}$ over A. Embed (A, L) into (C_1, C_2) as (A^*, L^*) (Lemma 20.3.40).

Take $c \in C_2$ and an automorphism α of Γ carrying (A, c) into (A^*, L). Similarly find an automorphism β carrying (B, c) into (A^*, L). Then α, β preserve C_1 and C_2 and $\alpha[A] = \beta[B]$. So there is an automorphism of $C_1 \cup C_2$ taking A onto B.

The claim follows. □

20.3.5. Generic type: structure of u^2.

LEMMA 20.3.42. *Let Γ be an imprimitive homogeneous 2-multi-tournament of generic type with components of type T.*
Then for $u \in \Gamma$, u^2 is not of shuffled type.

PROOF. By Lemma 20.3.30 the components of u^2 are of type T. Thus it suffices to consider the cases $T = \mathbb{Q}$ or $\mathbb{S}$.

Let $L = \{u_1, u_2, u_3, u_4\}$ be transitive of order 4 with

$$u_1 \xrightarrow{1} u_2 \xrightarrow{1} u_3 \xrightarrow{1} u_4.$$

Take u, a_1, a_2 satisfying

$$u \xrightarrow{2} u_1, u_2, u_4, \quad u_3 \xrightarrow{2} u, \quad u \xrightarrow{1} a_1 \qquad La_1 \cong Lu;$$

$$a_2 \xrightarrow{2} u_1, u_3, u_4, \quad u_2 \xrightarrow{2} a_2 \quad u, \xrightarrow{2} a_2 \text{ or } a_2 \xrightarrow{2} u.$$

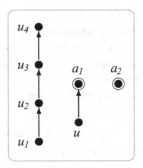

In an amalgam of Lua_1 with Lua_2 over Lu, we have $a_1 \xrightarrow{2} a_2$ or $a_2 \xrightarrow{2} a_1$ and correspondingly $a_1 \xrightarrow{2} (u_1 u_2 u_4 a_2)$ or $a_2 \xrightarrow{2} (u_1 u_3 u_4 a_1)$. It then follows that a_1' or a_2' is not of shuffled type. So it suffices to show that the factors omitting a_2 or a_1 in this diagram embed into Γ.

The factor omitting a_2 has only two $\overset{1}{\sim}$-classes and is afforded by Lemma 20.3.41.

For the factor omitting a_1 it suffices to realize the types of u and of a_2 over L in distinct $\overset{1}{\sim}$-classes. Again by Lemma 20.3.41, after embedding L into one such class, each of these types may be realized in any other class. $\square$

LEMMA 20.3.43. *Let Γ be an imprimitive homogeneous 2-multi-tournament of generic type and components of type T. Then for $u \in \Gamma$, u^2 is of generic type with components of type T.*

PROOF. By Lemma 20.3.30 the components of u^2 are of type T. Since u^2 is not a composition or shuffled, it must be semi-generic or of generic type. So suppose toward a contradiction that u^2 is semi-generic. In particular, any configuration satisfying the parity constraint embeds into Γ.

Consider the following amalgamation.

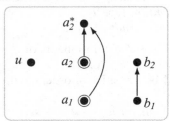

$$u \xrightarrow{2} a_1, a_2, b_1, b_2, \qquad\qquad a_2^* \xrightarrow{2} u;$$

$$a_1 \xrightarrow{2} b_1, b_2, \qquad\qquad a_2, a_2^* \xrightarrow{2} b_1, \qquad b_2 \xrightarrow{2} a_2, a_2^*.$$

In any amalgam a_1, a_2, b_1, b_2 will violate the parity constraint and lie in u^2. So it suffices to check that the factors omitting a_1 or a_2 embed in Γ.

As a_2, a_2^* realize the same type over b_1, b_2, the parity constraint holds in the factor omitting a_1.

The factor omitting a_2 is obtained by amalgamating the subfactors $ua_2b_1b_2$ with $a_2a_2^*b_1b_2$ to determine the type of (u, a_2^*). If $a_2^* \xrightarrow{2} u$ we have the required factor and if $u \xrightarrow{2} a_2^*$ we violate the parity constraint in u^2.

Finally, the subfactors $ua_2b_1b_2$ and $a_2a_2^*b_1b_2$ embed into Γ as both satisfy the parity constraint.   □

20.3.6. Infinite quotient, generic type: conclusion.

NOTATION 20.3.44. Suppose that Γ is a homogeneous 2-multi-tournament of generic type with $\Gamma/\overset{1}{\sim}$ infinite, with components of type T, and with associated amalgamation class $\mathcal{A}$. Let $\mathcal{A}^*$ be the class of all $A \in \mathcal{A}$ such that

for all finite expansions $A \cup B$ of A by one additional $\overset{1}{\sim}$-class B which embeds in T, we have $AB \in \mathcal{A}$.

Notice that $\mathcal{A}^*$ is again an amalgamation class. Let Γ^* denote the corresponding homogeneous 2-multi-tournament. By Lemma 20.3.41, this has the same components as Γ.

A word of explanation: our expectation here is that $\mathcal{A}^* = \mathcal{A}$, and if we knew that this is the case, then the identification of Γ would be trivial: it would follow by induction on the number of $\overset{1}{\sim}$-classes that a finite imprimitive structure with components embedding in T must itself embed into Γ. The way we actually argue is to show that $\mathcal{A}^*$ is sufficiently like $\mathcal{A}$ to make this argument work. This idea of induction over amalgamation classes goes back to the classification of homogeneous graphs by Lachlan and Woodrow (and was key also in Part 1).

LEMMA 20.3.45. *Let Γ be an imprimitive homogeneous 2-multi-tournament of generic type. Then the configuration $A = (a, b)$ with a $\xrightarrow{2}$ b lies in $\mathcal{A}^*$.*

PROOF. We make an amalgamation of the form $a_1uB_1B_2$ with $a_2uB_1B_2$ over uB_1B_2 with $u \xrightarrow{1} a_1$ and $u \xrightarrow{2} a_2$ so that $(a_1a_2B_1)$ or $(a_2a_1B_2)$ is isomorphic to AB. The element u ensures that $a_1 \xrightarrow{2} a_2$ or $a_2 \xrightarrow{2} a_1$. We also take $u \xrightarrow{2} B_1B_2$.

The factor omitting a_2 has the two $\overset{1}{\sim}$-classes (u, a_1) and B_1B_2. The factor omitting a_1 consists of $a_2B_1B_2$ in u^2. As u^2 is again of generic type this factor embeds in Γ as well.   □

Now we show that Γ^* is not composite.

LEMMA 20.3.46. *Let Γ be an imprimitive homogeneous 2-multi-tournament of generic type. Then the configuration $A = (a_1, a_2, b)$ with $a_1 \xrightarrow{1} a_2$ and $a_1 \xrightarrow{2} b \xrightarrow{2} a_2$ lies in $\mathcal{A}^*$.*

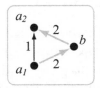

FIGURE 4. A (Lemma 20.3.46).

PROOF. We consider an extension AB by an additional $\sim$-class. We set up an amalgamation as follows to force $(a_1 a_2 u B_1)$ or $(a_2 a_1 v B_2)$ to be isomorphic to the given extension AB.

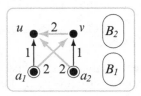

Now the factor omitting a_1 is of the form $A_2 B_1 B_2$ with $A_2 = (u, v, a_2)$. Here $A_2 \in \mathcal{A}^*$ since $\mathcal{A}^*$ is an amalgamation class containing all finite transitive $\xrightarrow{1}$-tournaments as well as the configuration of Lemma 20.3.45. Thus it suffices to deal with the factor $(a_1 u v B_1 B_2)$.

We may treat this factor as an amalgamation problem in which the type of $u B_2$ is to be determined. Its factors are $(a_1 v B_1 B_2)$ and $(a_1 u v B_1)$. The former we already know is embedded in Γ (Lemma 20.3.45).

For the factor $(a_1 u v B_1)$, we fix some $b \in B_1$ and treat the configuration as an amalgamation problem to determine the type of v over $B_1 \setminus \{b\}$. Again, the factor $(a_1 u B_1)$ presents no problem and we come down to the factor $(a_1 u v b)$. But this may be written as $(vb) \cup \{a_1, u\}$ and again Lemma 20.3.45 applies.   □

Thus Γ^* is not composite. Continuing in the same vein, we show next that Γ^* is not of shuffled type.

LEMMA 20.3.47. *Let Γ be an imprimitive homogeneous 2-multi-tournament of generic type. Then the configuration $A = (a_1, a_2, a_3, b)$ with $a_1 \xrightarrow{1} a_2 \xrightarrow{1} a_3$ transitive and $a_2 \xrightarrow{2} b \xrightarrow{2} a_1, a_3$ lies in $\mathcal{A}^*$.*

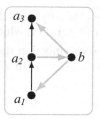

PROOF. We consider any extension AB of the desired type, and make the following amalgamation, designed to force $(u_1 a_1 u_3 a_2 B_1)$ or $(a_1 u_2 u_3 a_2 B_2)$ to be isomorphic to AB, according as $a_1 \xrightarrow{2} a_2$ or $a_2 \xrightarrow{2} a_1$.

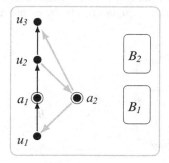

It suffices to show that the factors embed in Γ.

The factor omitting a_2 is in Γ by Lemma 20.3.41. The factor omitting a_1 is obtained by treating it as an amalgamation problem, to determine the type of u_1 over B_2.

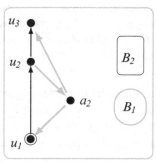

In this amalgamation, the factor omitting u_1 is covered by Lemma 20.3.46. The factor omitting B_1 is constructed as an amalgamation determining the type of u_2 over B_1. Both factors of this last diagram embed in Γ.   □

LEMMA 20.3.48. *Let Γ be an imprimitive homogeneous 2-multi-tournament of generic type. Then the associated homogeneous 2-multi-tournament Γ^* violates the parity constraint.*

Proof. Assume the contrary. Then by our results so far, Γ^* is semi-generic. In particular Γ contains any configuration which does not violate the parity constraint, as long as the $\overset{1}{\sim}$-classes embed in the components of Γ.

Let A be the configuration (L_1, L_2) with $L_1 = \{a_1, a_2\}$, $L_2 = \{b_1, b_2\}$, satisfying

$$a_1 \xrightarrow{1} a_2, \qquad\qquad\qquad b_1 \xrightarrow{1} b_2,$$
$$a_1 \xrightarrow{2} b_1, b_2, \qquad\qquad\qquad b_2 \xrightarrow{2} a_2 \xrightarrow{2} b_1.$$

It suffices to show that A belongs to $\mathcal{A}^*$. So we fix an extension AB of the desired type and consider the following amalgamation diagram, designed to force one of $(u_1 a_1 v a_2 B_1)$ or $(a_1 u_2 v a_2 B_2)$ to be isomorphic to AB.

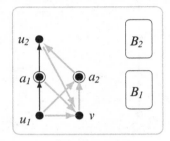

The configuration $(u_1 a_1 u_2 v)$ satisfies the parity constraint, vacuously, and hence belongs to $\mathcal{A}^*$ under our current hypotheses. So the factor $(u_1 a_1 u_2 v B_1 B_2)$ embeds in Γ.

The factor $(u_1 u_2 v a_2 B_1 B_2)$ may be viewed as an amalgamation diagram determining the type of u_2 over B_1. Again $(u_1 v a_2)$ belongs to $\mathcal{A}^*$ so we come down to the factor $(u_1 u_2 v a_2 B_2)$. This may be viewed as an amalgamation diagram determining the type of u_1 over B_2, with factors $(u_1 u_2 v a_2)$ and $u_2 v a_2 B_2$, both embedding in Γ. $\qquad\qquad\square$

Now we may conclude.

Lemma 20.3.49. *Let Γ be an imprimitive homogeneous 2-multi-tournament of generic type with components of type T. Then Γ is the generic imprimitive 2-multi-tournament with components of type T.*

Proof. We show the following by a very formal induction on n.

> Any finite 2-multi-tournament K with at most n $\overset{1}{\sim}$-classes whose components embed into T embeds into Γ. $\qquad (*_n)$

Furthermore we construe this as applying, with n fixed, to *any* imprimitive homogeneous 2-multi-tournament of generic type.

We know this property already for $n = 2$, so we suppose that $n \geq 3$ and the claim holds for fewer than n $\overset{1}{\sim}$-classes.

Claim 1. *Any finite 2-multi-tournament K of order $(n-1)$ whose components embed into T lies in $\mathcal{A}^*$.*

We consider a configuration KB with B an additional $\overset{1}{\sim}$-class. If K has at most $(n-2)$ $\overset{1}{\sim}$-classes then KB embeds into Γ by assumption. So we may suppose the verticees of K lie in distinct $\overset{1}{\sim}$-classes. Fix $a, b \in K$ distinct, and let $K_0 = K \setminus \{a, b\}$. Form the usual amalgamation diagram $u a_1 a_2 K_1 K_2 B_1 B_2$ so as to force $(a_1 a_2 K_1 B_1)$ or $(a_2 a_1 K_2 B_2)$ to be isomorphic to KB, with

$$u \overset{1}{\longrightarrow} a_1, \qquad\qquad u \overset{2}{\longrightarrow} a_2 K_1 K_2 B_1 B_2.$$

It suffices to check that the factors omitting a_2 or a_1 embed into Γ.

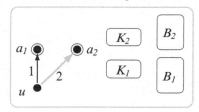

The factor omitting a_2 has only $(n-1)$ $\overset{1}{\sim}$-classes, so embeds in Γ by hypothesis. The factor omitting a_1 has the form $(u \overset{2}{\longrightarrow} a_1 K_1 K_2 B_1 B_2)$ and u^2 is of generic type, so by the induction hypothesis, applied to u^2, this factor also embeds into Γ.

The claim follows.

It follows that Γ^* is of general type with at least $(n-1)$ classes, and does not satisfy the parity constraint. If Γ^* has finitely many classes then it is of generic type by our previous analysis. If Γ^* has infinitely many classes then our induction hypothesis applies and Γ^* contains any configuration with $(n-1)$ $\overset{1}{\sim}$-classes whose components embed into T.

Hence in any case, if A is a configuration with $(n-1)$ $\overset{1}{\sim}$-classes whose components embed into T then A is in $\mathcal{A}^*$, and hence AB embeds in Γ, as required.

Thus condition $(*_n)$ holds for all n, and Γ is generic.   □

At this point, as we now verify, the classification of the imprimitive homogeneous 2-multi-tournaments is complete.

Proof of Proposition 20.3.2. By Lemma 20.3.14, Γ is of shuffled type or general type.

The classification of shuffled type is completed in Lemma 20.3.22.

For general type with $\Gamma/\overset{1}{\sim}$ finite, the classification is given in Lemma 20.3.34. When $\Gamma/\overset{1}{\sim}$ is infinite, the classification is given in the semi-generic case in Lemma 20.3.36 and in the generic case in Lemma 20.3.49.   □

According to the general philosophy, or expectation, that the 3-constrained structures are the key to classifying homogeneous structures for finite binary languages, the next point to consider is the classification of the 3-constrained cases. We deal with this point in the next chapter.

Chapter 21

3-CONSTRAINED HOMOGENEOUS
2-MULTI-TOURNAMENTS

With the classification of the imprimitive homogeneous 2-multi-tournaments in hand, we turn to the primitive case. We take up the following two points, which make up the first phase of a systematic analysis.

(a) What are the 3-constrained homogeneous 2-multi-tournaments?
(b) Must every homogeneous 2-multi-tournament not in our catalog have a pattern of forbidden triangles corresponding to some 3-constrained amalgamation class?

(Cf. §18.3.6.)

Recall that the 3-constrained homogeneous structures are those for which the associated amalgamation class is determined by a class of forbidden structures of order at most 3. In the case of 2-multi-tournaments their classification was dealt with very briefly, and without proof, in the appendix to Cherlin [1998].

We will prove here that the table given in that reference is correct. This involves three points.

(I) Demonstrating the existence of these structures;
(II) Giving natural interpretations of these structures, when we are aware of them;
(III) Proving the completeness of the list given.

This last point—the full classification of these structures—requires amalgamation arguments which are very much simplified by the assumption that the structure is 3-constrained.

Conjecturally, essentially the same patterns of forbidden triangles should be associated with any homogeneous 2-multi-tournament, with known exceptions in the finite or imprimitive cases. This is our second problem, a much harder one. We take this up in the next chapter, and we do not fully resolve it.

Later we will see that in many cases the easy amalgamation arguments of the present chapter can be replaced by more elaborate ones leading to the same results. If the structure is 4-*trivial* in the sense that every configuration of order 4 whose triangles embed in Γ will itself embed in Γ, then most of the argument would run along the same lines as in the 3-constrained case. Occasionally instances of 5-triviality are also required. We can work our way around the need for 4-triviality but where 5-triviality is needed more severe difficulties arise.

We remind the reader that a catalog of all known homogeneous 2-multi-tournaments is presented in §20.2, and that this catalog is complete as far as the degenerate, finite, or imprimitive cases are concerned (Fact 20.2.3, Proposition 20.3.2). A more technical version of the catalog, in terms of the minimal forbidden substructures, will be given in our final Tables 22.1 (imprimitive case) and 22.2 (primitive, non-free case).

Furthermore, in Part I we classified the homogeneous 2-multi-tournaments with a definable linear order. These were referred to in Part I, interchangeably—for our purposes—as ordered graphs or as ordered tournaments, as discussed in Chapter 1 of Volume 1, §1.3.

21.1. Statement of the classification

We aim at the following.

PROPOSITION 21.1.1. *Up to a permutation of the language, the primitive non-degenerate 3-constrained homogeneous 2-multi-tournaments are the twelve shown in Table 18.1 together with the four free amalgamation classes which forbid only* $C_3(1, 1, 1)$, *or* $L_3(1, 1, 1)$, *or neither, or both.*

Since we deal with two asymmetric 2-types, the permutations of the language considered here may switch the two labels and reverse either or both 2-types.

The list given in Table 18.1 includes a finite 2-multi-tournament and four infinite ones having definable linear orders. Since the finite case and the case in which there is a definable linear order are also covered by complete classification results, at this point Proposition 21.1.1 reduces to the following (cf. §18.3.6.1).

PROPOSITION 21.1.2. *Up to a permutation of the language, the primitive non-degenerate 3-constrained homogeneous 2-multi-tournaments with no definable linear ordering are the eleven shown in Table 21.1.*

The entries in Table 21.1 are numbered 6 through 16. There entries 6 through 12 are the last seven entries of Table 18.1, and the additional entries are the free amalgamation classes. These structures are divided into groups III–V with groups III and IV taken over from Table 18.1 and with group V consisting of the free amalgamation classes.

We will now discuss the classification as shown Table 21.1 in detail. In particular we review the notation used for constraints, and the "Type" descriptions of the individual cases.

The notation for constraints of order three was established in Notation 18.3.1, using the symbols C_3, L_3 for 3-cycles and transitive tournaments of order three,

Primitive infinite 3-constrained homogeneous		
2-multi-tournaments with no $\emptyset$-definable linear order		
#	*Constraints*	*Type*
Group III: p.o. by $\xrightarrow{1}$, no $\emptyset$-definable linear order		
Common Constraints, entries 6 and 7:		
C_3: 111, 112 L_3: 112		
Additional Constraints, entries 6 and 7:		
6	C_3: none L_3: none	$\widetilde{\mathcal{P}}$
7	C_3: 221 L_3: none	Exceptional?
Group IV: Infinite, no $\emptyset$-definable partial order, not free		
Constraints		
8	C_3: 111,112 L_3: none	Exceptional?
9	C_3: 112 L_3: 111	"
10	C_3: 111,112 L_3: 111	"
11	C_3: 111,222 L_3: 111,122,212	$\widetilde{\mathbb{S}(3)}$
12	C_3: 111,112 L_3: 121,211,221,222	$\mathbb{S}(4)$
Group V: Free amalgamation classes		
Constraints		
13–16	C_3: 111 or none L_3: 111 or none	Free

TABLE 21.1. Primitive infinite 3-constrained homogeneous 2-multi-tournaments with no linear order.

respectively, and the labels $1, 2$ for the arc colors (2-types). In particular the free amalgamation classes (group V) are those with monochromatic constraints of one "color" only; we may suppose that color is 1.

The main distinction in the table is between group III, where we have a definable partial order (but, by hypothesis, no definable linear order), and group IV, where we have no definable partial order, and we do not have free amalgamation either.

In group III we may suppose the definable (strict) partial order coincides with $\xrightarrow{1}$. In other words the relation $\xrightarrow{1}$ is transitive. This condition corresponds to the three forbidden triangles listed as "Common Constraints" in this case. Further constraints are listed individually by case (except in group V where we have bundled the free cases together, in an unspecified order, as we will not be examining them in any detail).

Now we come to the entry under "Type." Of the seven interesting examples (numbered 6–12) we have sensible descriptions of three, so we abandon the other four to the "exceptional" category.[4] We will see however that the structures 8–10 have a weak kind of free amalgamation, and an associated notion of Henson constraint, giving rise to a larger family of homogeneous structures which are not 3-constrained, but are standard in the sense of Definition 18.2.1.

The three structures that we do understand in a direct way are labeled as follows.

1. $\mathcal{P}$ (without the tilde) is the generic partial order; this is a homogeneous directed graph.
2. The structures $\mathbb{S}(n)$ will be introduced in Definition 21.2.7. $\mathbb{S}(2)$ is $\mathbb{S}$, $\mathbb{S}(3)$ is a homogeneous directed graph we have called the *myopic local order*, and $\mathbb{S}(4)$ is a homogeneous 2-multi-tournament. (The complexity of the language increases with n.)
3. The tilde is used to denote the operation of generic de-symmetrization, and specifically, the operation which takes as input a homogeneous directed graph whose associated amalgamation class has strong amalgamation, and constructs the corresponding homogeneous 2-multi-tournament in which the symmetric non-edge relation of the directed graph is generically split into an arc relation and its reverse.

Putting these conventions together we have the following identifications.

Entry #6 is the generically de-symmetrized generic partial order;
Entry #11 is the generically de-symmetrized myopic local order;
Entry #12 is $\mathbb{S}(4)$; here $\mathbb{S}(n)$ is a homogeneous structure for a binary language, and for $n = 4$ it is a 2-multi-tournament.

Entry #7 is a non-generic de-symmetrization of the generic partial order, in which an additional constraint is imposed. In the context of symmetric languages, we expect the additional constraints to be Henson constraints in the sense that they involve types not needed for amalgamation in the larger class; but it is not clear how one might adapt this point of view to the anti-symmetric case.

This explains the meaning of our table. We claim that the listed constraints define amalgamation classes, so that their Fraïssé limits exist and are homogeneous 2-multi-tournaments; that the resulting structures are as described, when we have a meaningful description; and that this exhausts the class of infinite primitive non-degenerate 3-constrained homogeneous 2-multi-tournaments with no definable linear order.

---

[4]Actually, the description of these structures in terms of forbidden *partial substructures* in Lemmas 21.2.3 and 21.2.5 is more informative than the description in terms of forbidden substructures given in the table, and makes the amalgamation property relatively transparent.

Notation 21.1.3. It will be convenient to have a notation for monochromatic 2-multi-tournaments; that is, having only one of the arc relations $\overset{1}{\to}$ or $\overset{2}{\to}$. If T is a finite tournament (typically, with arc relation denoted $\to$), then for $i = 1$ or 2 we let T^i be the monochromatic 2-multi-tournament derived from T by replacing the symbol for its arc relation by the symbol $\overset{i}{\to}$. So only the language changes; but there are two possibilities and they need to be distinguished.

Occasionally we may use the variant notation $T(\overset{i}{\to})$ or even $T(i)$.

Note that this usage has nothing to do with the earlier Notation 20.1.1, A^i, extending the notation a^i from points to configurations. Fortunately that notation is not in use here.

21.2. Existence

The first point to take up is the following.

Proposition 21.2.1. *The classes specified by the various constraint sets shown in Table* 21.1 *are amalgamation classes.*

The cases which require attention are entries 7–10, where we check amalgamation directly, and entries 6, 11, 12, where we must justify the claim that the constraints shown characterize known structures (and, in the case of 11–12, first make them known).

We may dispose of entry #6 at once. Certainly the generic de-symmetrization of the generic partial order exists (that is, we desymmetrize the incomparability relation), and by definition it is a homogeneous 2-multi-tournament. It is characterized by the transitivity of the relation $\overset{1}{\to}$, which, as we have observed, amounts to forbidding the three triangles listed.

So it remains to treat the exceptional cases by checking the amalgamation property, and to define the structures $\mathbb{S}(n)$, and also determine their defining constraints.

21.2.1. The exceptional cases: 7–10. It is easy to check amalgamation for entry (7) by directly checking 2-point amalgamation; one takes as the completing type the type $\overset{1}{\to}$ only if this is required by transitivity, and otherwise one takes $\overset{2}{\to}$, taking some care to satisfy the constraints. However we prefer to prove something a little stronger, which is useful for other more combinatorial purposes, by identifying the constraints for the *partial substructures*.

Definition 21.2.2. A *partial 2-multi-tournament* is a structure which is equipped with two disjoint anti-symmetric binary relations $\overset{1}{\to}$ and $\overset{2}{\to}$.

We say that a partial 2-multi-tournament extends to a given 2-multi-tournament A if it embeds into A as a weak substructure (i.e., as a substructure in the graph theoretic sense), adding vertices and arcs.

LEMMA 21.2.3. *A partial 2-multi-tournament extends to some structure of type (7) iff it satisfies the following two conditions.*

1. *every oriented cycle contains at least three arcs of type $\xrightarrow{2}$.*
2. *The transitive closure of $\xrightarrow{1}$ does not contain a pair with type $\xrightarrow{2}$.*

PROOF. The necessity is straightforward: to see that oriented cycles with at most two occurrences of $\xrightarrow{2}$ are forbidden, apply the transitivity of $\xrightarrow{1}$ to reduce to consideration of cycles of length 3 or 4 with two occurrences of $\xrightarrow{2}$. The former are forbidden by assumption and the latter by inspection.

So we turn to sufficiency. Suppose we have a partial 2-multi-tournament $(X, \xrightarrow{1}, \xrightarrow{2})$ satisfying the stated conditions.

Then there is no non-trivial oriented closed walk (i.e., oriented cycle, but allowing self-intersections) with fewer than three arcs of type $\xrightarrow{2}$.

First replace $\xrightarrow{1}$ by its transitive closure. By our two assumptions, this extension is consistent with the structure already present. The first condition is maintained since any walk in the extended structure gives rise to a walk in the original structure with the same endpoints and the same number of occurrences of $\xrightarrow{2}$.

So we suppose that the relation $\xrightarrow{1}$ is transitive. As $\xrightarrow{1}$ will not be further extended, the condition (2) is settled definitively at this point.

Now we proceed by iterating the following algorithm. We will check that it preserves our constraints.

(I) As long as there is a pair of vertices a, b whose type is not already determined, for which there is an oriented path from a to b with exactly one occurrence of $\xrightarrow{2}$, set $a \xrightarrow{2} b$; By condition (1), the extended relation $\xrightarrow{2}$ is anti-symmetric.

(II) If this stabilizes, and the structure is not complete, select one undetermined pair (a, b) and set $a \xrightarrow{2} b$; return to (I).

It is necessary to check that condition (1) is preserved throughout. After an application of Step (I) any walk in the extended structure corresponds to a walk in the original structure with the same number of occurrences of the type $\xrightarrow{2}$. After an application of Step (II) any new cycle consists of the new arc $a \xrightarrow{2} b$ and a path in the previous structure; if our condition were violated we would in fact be in Step (I). □

LEMMA 21.2.4. *Entry (7) from Table 21.1 defines an amalgamation class.*

PROOF. Given an amalgamation problem $(A_0 \to A_1, A_2)$, we first amalgamate A_1, A_2 freely over A_0 to get a partial 2-multi-tournament $\hat{A}$. It suffices to check the two conditions of Lemma 21.2.3. We know that the transitive closure of $\xrightarrow{1}$ will not conflict with the structure on A_1 or on A_2; this is the usual amalgamation of partial orders.

Let γ be an oriented cycle lying in the free amalgam $\hat{A}$. Suppose that γ has at most two occurrences of $\xrightarrow{2}$, and that the length of γ is minimized.

If γ is contained in A_1 or in A_2 then we have a contradiction. So γ must contain two vertices u, v in A_0 which are not adjacent in γ. Then γ divides into paths γ_1 from u to v and γ_2 from v to u of length at least two. In particular the cycle (u, v, γ_2) is shorter than γ and thus has at least three arcs ot type 2; one of these must be (u, v) and then exactly two lie in γ_2. Hence γ_1 is a path of 1-arcs.

We may suppose that u, v are chosen in $\gamma \cap A_0$ so that the distance $d(u, v)$ measured along γ_1 is minimized, subject to our constraint that it is at least 2. Then γ_1 lies wholly in A_1, or wholly in A_2, and then condition (2) is violated in that factor. $\qquad\square$

LEMMA 21.2.5 (Structures #8–#10). *Each of the classes of finite 2-multi-tournaments determined by one of the following sets of forbidden triangles (numbered as in Table* 21.1) *is an amalgamation class, with a weak form of free amalgamation: any amalgamation problem can be completed using some combination of* $\xrightarrow{2}$ *and* $\xleftarrow{2}$; *in other words, without using either form of* $\xrightarrow{1}$.

(8) $C_3(1, 1, 1)$, $C_3(1, 1, 2)$.
(9) $C_3(1, 1, 2)$, $L_3(1, 1, 1)$.
(10) $C_3(1, 1, 1)$, $C_3(1, 1, 2)$, $L_3(1, 1, 1)$.

We note in passing that in this statement, case (10) is a special case of case (8) or (9), since it differs only by an additional monochromatic constraint of type 1. The proof we give is uniform across all three cases: what matters here is that we forbid $C_3(1, 1, 2)$ and some monochromatic triangle of type 1 (that is, at least one).

PROOF. We consider a 2-point amalgamation problem

$$(A_0 \cup \{a\}, A_0 \cup \{b\}).$$

The claim is that in any of the three cases we can complete the diagram by either $a \xrightarrow{2} b$ or $b \xrightarrow{2} a$ without creating a 3-cycle of type $C_3(1, 1, 2)$.

We can use $a \xrightarrow{2} b$ unless we have a configuration

$$b \xrightarrow{1} x \xrightarrow{1} a$$

with $x \in A_0$, and we can use $b \xrightarrow{2} a$ unless we have a configuration

$$a \xrightarrow{1} y \xrightarrow{1} b$$

with $y \in A_0$.

In other words, the claim holds unless we have a 4-cycle of type $\xrightarrow{1}$ of the form (a, x, b, y).

But in such a 4-cycle there would be no type available for the arc between x and y, by inspection.   □

Again, we can give a sharper version of the amalgamation property, in terms of a characterization of partial substructures. The content is very similar to the content of the preceding proof.

LEMMA 21.2.6 (#8–#10: Partial substructures). *For each of the classes of finite 2-multi-tournaments determined by one of the following sets of forbidden triangles (numbered as in Table* 21.1), *the corresponding class of partial 2-multi-tournaments is given by the same set of triangle constraints, plus an oriented 4-cycle of type* $\xrightarrow{1}$.

(8) $C_3(1, 1, 1)$, $C_3(1, 1, 2)$.
(9) $C_3(1, 1, 2)$, $L_3(1, 1, 1)$.
(10) $C_3(1, 1, 1)$, $C_3(1, 1, 2)$, $L_3(1, 1, 1)$.

This also gives amalgamation, but for much the same reason as originally given, now stated explicitly in terms of forbidden 4-cycles.

PROOF. *Necessity*: As noted in the previous proof, an oriented 4-cycle cannot be extended by even one more arc so as to meet these constraints.
Sufficiency: If we have a pair of vertices a, b with no arc assigned and an oriented path $a \xrightarrow{1} x \xrightarrow{1} b$ between them, take $a \xrightarrow{2} b$. This cannot introduce any monochromatic configuration of type $\xrightarrow{1}$ (in particular, no 4-cycle of this type) and it cannot introduce $C_3(1, 1, 2)$, since there is no oriented 4-cycle of this type already present.

Having dealt with all such cases, fill in any additional arcs needed as $\xrightarrow{2}$ with an arbitrary orientation.   □

It is seems striking that the completion process is the same in all three cases 8–10. We remark that in the symmetric case, canonical completion processes tend to be variations on shortest path metrics (for generalized metric spaces). It would be interesting to explore of the 3-constrained structures and the completion algorithms for partial substructures in richer languages with anti-symmetric relations, but this lies well outside the scope of this volume.

Since we can amalgamate using only some form of the type $\xrightarrow{2}$ we have an associated notion of Henson constraint. Namely, we can take the constraints in case 8—$C_3(1, 1, 1)$ and $C_3(1, 1, 2)$, and in addition forbid L^1_{n+1} for some fixed value of $n \geq 2$ (recall Notation 21.1.3).

For $n = 2$ we get entry #10, while for larger values of n, or no additional constraint, we get the same forbidden triangles as in entry #8. We will refer to this case later as $3C_{8:n}$.

21.2.2. The structures $\mathbb{S}(n)$ and entries 11 and 12. The issue with regard to entries #11 and #12 is not so much the question of existence (amalgamation) as the identification of these structures with structures already known.

DEFINITION 21.2.7. The structure $\mathbb{S}(n)$ is defined as follows, for $n \geq 2$. The points are the image of $\alpha\mathbb{Q}$ in $\mathbb{R}/n\mathbb{Z}$ where α is some fixed irrational. For a point a and $i \in \mathbb{Z}/n\mathbb{Z}$ set $\mathrm{ind}(a) = i$ if $a - i$ has a representative in $(0, 1)$. We may call this the *index* of a in $\mathbb{Z}/n\mathbb{Z}$; it will not be part of the language but will help with the definition of 2-types. Namely, the type of a pair of distinct elements $a, b \in \mathbb{S}(n)$ is taken to be

$$\mathrm{ind}(b - a)$$

and if this is i, we also write $a \xrightarrow{i} b$. Then $\mathbb{S}(n)$ consists of the specified points together with the relations $a \xrightarrow{i} b$.

We will show that for $n \geq 3$ the structure $\mathbb{S}(n)$ is 3-constrained and homogeneous (Lemma 21.2.11). For $n = 2$ it is also homogeneous—$\mathbb{S}(2) = \mathbb{S}$ is the generic local order—but not 3-constrained, and in fact there are no constraints on triangles. This is an interesting example of a primitive structure in a finite binary language not derived from a 3-constrained amalgamation class by imposing Henson constraints. We were surprised to notice that the natural generalization of $\mathbb{S}$ is so much tamer for $n \geq 3$.

Parts (1,4) of the next lemma suggest that $\mathbb{S}(n)$ has some affinity with generalized metric spaces with not necessarily symmetric distance. Writing i^* for $-i - 1$ and $x +^* y$ for $x + y + 1$, we have the laws $d(b, a) = d(a, b)^*$, $(i^* + j^*)^* = i +^* j$, and $d(x, z) = d(x, y) + d(y, z)$ or $d(x, y) +^* d(y, z)$. See also Appendix A, Vol. I.

LEMMA 21.2.8. $\mathbb{S}(n)$ *has the following properties.*

1. *The relation $a \xrightarrow{i} b$ coincides with the relation $b \xrightarrow{-i-1} a$.*
2. *If n is even then $\mathbb{S}(n)$ is an $n/2$-multi-tournament. If n is odd then $\mathbb{S}(n)$ has $(n-1)/2$ pairs of asymmetric 2-types and one symmetric 2-type. In particular, $\mathbb{S}(3)$ is a directed graph and $\mathbb{S}(4)$ is a 2-multi-tournament.*
3. *If $a, b \in \mathbb{S}(n)$ have index i, j respectively, then $\mathrm{tp}(a, b)$ will be $j - i$ or $j - i - 1$.*
4. *For $a, b, c \in \mathbb{S}(n)$ with $a \xrightarrow{i} b$, $b \xrightarrow{j} c$, $a \xrightarrow{k} c$ we have k equal to $i + j$ or $i + j + 1$, and both possibilities occur.*

PROOF. We treat each point in turn.

Ad 1. This simply means that $\mathrm{ind}(-x) = -\mathrm{ind}(x) - 1$, which is clear.

Ad 2. This follows from (1).

Ad 3. In view of (1) we may suppose $i \leq j$. Then $b - a$ has a representative in the set of differences $(j, j + 1) - (i, i + 1) \subseteq (j - (i + 1), j + 1 - i)$, and the claim follows.

Ad 4. As $c - a = (c - b) + (b - a)$ the condition on k follows from

$$(i, i + 1) + (j, j + 1) = (i + j, i + j + 2).$$

It is easy to see by the same analysis that both possibilities for k occur.   □

The next step in our analysis of $\mathbb{S}(n)$ involves fixing a point and describing the resulting structure in a modified language. This needs to be done at a somewhat greater level of generality, for which we give a definition.

DEFINITION 21.2.9. A structure in the language of $\mathbb{S}(n)$ is $\mathbb{S}(n)$-*like* if it satisfies the following three conditions.

(a) For each pair of distinct elements a, b exactly one of the relations $a \xrightarrow{i} b$ holds.
(b) The reversal of the relation $\xrightarrow{i}$ is the relation $\xrightarrow{-i-1}$.
(c) For any triangle (a, b, c) with $a \xrightarrow{i} b, b \xrightarrow{j} c, a \xrightarrow{k} c$, we have

$$k = i + j \text{ or } i + j + 1.$$

We have shown above that $\mathbb{S}(n)$ is $\mathbb{S}(n)$-like.

LEMMA 21.2.10. *Suppose that $n \geq 3$ and S is an $\mathbb{S}(n)$-like structure. Fix a point $v_* \in S$ and define $x < y$ on $S' = S \setminus \{v_*\}$ by*

$$\exists i, j \in \mathbb{Z}/n\mathbb{Z} \text{ such that } v_* \xrightarrow{i} x, v_* \xrightarrow{j} y, x \xrightarrow{j-i} y.$$

Then $<$ is a linear order on S'.

PROOF.

CLAIM 1. $<$ *is a tournament on S'.*

Let a, b be distinct elements of S' with $v_* \xrightarrow{i} a, v_* \xrightarrow{j} b, a \xrightarrow{k} b$. Then $b \xrightarrow{-k-1} a$. We have $k = j - i$ or $j - i - 1$, and $-k - 1 = i - j - 1$ or $i - j$ correspondingly. So $<$ is a tournament.

CLAIM 2. $<$ *is transitive.*

We take a, b, c distinct with $v_* \xrightarrow{i} a, v_* \xrightarrow{j} b, v_* \xrightarrow{k} c$ and $a \xrightarrow{j-i} b$, $b \xrightarrow{k-j} c$.

As S is $\mathbb{S}(n)$-like, by considering (a, b, c) we find that $a \xrightarrow{k-i} c$ or $a \xrightarrow{k-i+1} c$. Suppose toward a contradiction the latter occurs. Then we have $v_* \xrightarrow{i} a \xrightarrow{k-i+1} c$ and thus $v_* \xrightarrow{k+1} c$ or $v_* \xrightarrow{k+2} c$ as S is $\mathbb{S}(n)$-like. So $k = k + 1$ or $k + 2$. But $n \geq 3$, so this is a contradiction.   □

It is well known that this construction also works for $\mathbb{S}(2)$, which is the generic local order, but for $n = 2$ every tournament is $\mathbb{S}(2)$-like in the sense of our definition, and in this case the construction requires additional constraints on configurations of order 4.

LEMMA 21.2.11. *For $n \geq 3$ the structure $\mathbb{S}(n)$ is a 3-constrained homogeneous structure. Furthermore, it is the only homogeneous structure for its language having the specified constraints on triangles.*

PROOF. It is easily checked that the construction of the previous lemma, when applied to $\mathbb{S}(n)$ and any chosen point a, produces the generic n-partitioned linear order: that is, a copy of $(\mathbb{Q}, <)$ partitioned into n dense classes. Since the latter is homogeneous for a language with unary predicates for the classes and the order relation, it follows that $\mathbb{S}(n)$ is homogeneous (this is the usual line of argument).

Since the same construction applies to any finite $\mathbb{S}(n)$-like structure S, after embedding the latter into the generic n-partitioned linear order, this gives an embedding of S into $\mathbb{S}(n)$. It follows that $\mathbb{S}(n)$ is 3-constrained.

For the final statement, one has to see that the argument in the first paragraph applies to any homogeneous structure Γ in the language of $\mathbb{S}(n)$ having the same constraints on triangles. For this, we fix a point $a \in \Gamma$ and analyze the associated structure with n components, each of which is a non-trivial homogeneous linear order, hence a copy of $(\mathbb{Q}, <)$. One has to see that each of the classes is dense. But each point in a given class determines a non-trivial splitting of the points in any other class into two intervals, so the claim follows. (The rest is essentially due to Skolem [1920], generalizing Cantor.)   □

LEMMA 21.2.12. *The 3-constrained structures described in entries #11 and #12 of Table 21.1 are* $\mathbb{S}(3)$ *and* $\mathbb{S}(4)$ *respectively.*

PROOF. It suffices to check the constraints after making a suitable identification of the language.

Case 1. The structure $\mathbb{S}(3)$ *and its generic de-symmetrization.*

In its natural language, the structure $\mathbb{S}(3)$ has the 2-types $\xrightarrow{0}$, $\xleftarrow{0}$ (which is $\xrightarrow{2}$) and the symmetric relation $\xrightarrow{1}$. The forbidden triangle types, written in the order

$$(\mathrm{tp}(a, b), \mathrm{tp}(b, c), \mathrm{tp}(a, c))$$

are $(i, j, i + j - 1)$, where we may take $i \leq j$ and write 0^{op} for 2, giving the following.

$$(0, 0, 0^{\mathrm{op}}), \quad (1, 0^{\mathrm{op}}, 0^{\mathrm{op}}), \quad (0, 0^{\mathrm{op}}, 1), \quad (1, 1, 1),$$
$$(0^{\mathrm{op}}, 0, {}^{\mathrm{op}}, 0), \quad (0, 1, 0),$$

where the second row shows duplicates (up to the order of the vertices).

Now we relabel the types $0, 1$ as $2, 1$ and furthermore take the type 1 to be anti-symmetric (replacing the original amalgamation class by its de-symmetrization).

Our constraints are then as follows, omitting the duplication; but now the last triangle splits into two forms.

$$C_3(2, 2, 2) \quad L_3(2, 1, 2) \quad L_3(1, 2, 2) \quad C_3(1, 1, 1)$$
$$L_3(1, 1, 1)$$

which agrees with our previous description of this structure (#11).

Case 2. *The structure* $\mathbb{S}(4)$. Now we have the 2-types $0, 1, 2, 3 = 0, 1, 1^{\mathrm{op}}, 0^{\mathrm{op}}$ and the forbidden triangle types (i, j, k) with $k \neq i + j, i + j + 1$. As there (i, j, k) is the same as $(-j - 1, -i - 1, -k - 1)$ we eliminate those duplicates and show the rest (where we also take $i \leq j$ as elements of $\{0, 1, 2, 3\}$). We use labels in $\mathbb{Z}/n\mathbb{Z}$ to make the initial calculation clearer.

$$(0, 0, 2) \quad (0, 0, 3) \quad (0, 1, 0) \quad (0, 1, 3) \quad (0, 2, 0) \quad (0, 2, 1)$$
$$(0, 3, 1) \quad (1, 1, 0) \quad (1, 1, 1) \quad (1, 2, 1)$$

In terms of the types $0, 1$, after eliminating the remaining duplicates this translates, in order, to the configurations $C_3(0, 0, 1)$, $C_3(0, 0, 0)$, $L_3(0, 1, 0)$, $L_3(1, 1, 0)$, $L_3(1, 0, 0)$, $L_3(1, 1, 1)$. If we replace the labels $0, 1$ by $1, 2$ respectively we get the constraints corresponding to entry #12, as shown.

$$C_3(1, 1, 2) \quad C_3(1, 1, 1)$$
$$L_3(2, 2, 2) \quad L_3(2, 2, 1) \quad L_3(2, 1, 1) \quad L_3(1, 2, 1) \qquad \square$$

PROOF OF PROPOSITION 21.2.1. In view of Lemma 21.2.12 the only cases not identified with a known homogeneous 2-multi-tournament are entries #7–10, and in these cases amalgamation is proved in Lemmas 21.2.4 and 21.2.5. $\square$

21.3. Completeness: the partially ordered case

Having dealt with the existence of the 3-constrained homogeneous 2-multi-tournaments given in Table 21.1, we still need to argue that this list is complete: that is, that every primitive infinite 3-constrained homogeneous 2-multi-tournament with no $\emptyset$-definable linear order appears on this list.

This may be stated as follows.

PROPOSITION 21.3.1. *Let* Γ *be an infinite primitive non-degenerate homogeneous 3-constrained 2-multi-tournament on which there is no* $\emptyset$*-definable linear order. Then the set of forbidden triangles corresponds to one of the entries in Table* 21.1.

In more detail:

1. *If the type* $\xrightarrow{1}$ *is transitive, then all triangle types are realized which are compatible with that assumption, except, possibly, one of the pair* $L_3(2, 2, 1)$, $C_3(2, 2, 1)$.
2. *If there is no transitive 2-type, then the following hold.*
   (a) *If triangle types* $C_3(1, 1, 1)$ *and* $C_3(2, 2, 2)$ *are forbidden, then up to a change of language the structure is* $\mathbb{S}(3)$.
   (b) *If triangle type* $C_3(1, 1, 1)$ *is forbidden and triangle type* $C_3(2, 2, 2)$ *is realized, then up to a change of language* Γ *is determined by the constraints given in one of entries* #8 *or* #10 *of our table, or* Γ *is* $\mathbb{S}(4)$.

(c) *If both triangle types $C_3(1, 1, 1)$ and $C_3(2, 2, 2)$ are realized, then up to a change of language Γ is determined by the constraints in entry #9 of our table.*

Some of this may be proved more generally with approximately the same amount of effort. We phrase the first case as follows.

Proposition 21.3.2. *Let Γ be a primitive non-degenerate homogeneous 2-multi-tournament on which there is no $\emptyset$-definable linear order. Suppose the type $\xrightarrow{1}$ is transitive, that is,*

$$L_3(1, 1, 2), \ C_3(1, 1, 1), \text{ and } C_3(1, 1, 2) \text{ are forbidden.}$$

Then the following triangle types are realized

$$L_3(1, 1, 1) \qquad L_3(1, 2, 1) \qquad L_3(1, 2, 2) \qquad L_3(2, 1, 1) \qquad L_3(2, 1, 2)$$
$$L_3(2, 2, 2) \qquad C_3(2, 2, 2).$$

If $(IC_3)^2$ or $(C_3I)^2$ is realized in Γ then at least one of the triangle types $L_3(2, 2, 1)$ or $C_3(2, 2, 1)$ is realized as well; this applies in particular if Γ is 3-constrained.

Here IC_3 denotes the tournament consisting a point dominating an oriented 3-cycle, C_3I is its reversal, and in accordance with Notation 21.1.3, $(IC_3)^2$ or $(C_3I)^2$ denotes the same tournament with the arcs regarded as 2-arcs.

Remark 21.3.3. In principle under the stated conditions Proposition 21.3.2 determines the triangle constraints on Γ, up to the three possibilities corresponding to realizing at least one of $L_3(2, 2, 1)$ or $C_3(2, 2, 1)$.

However, the triangle type $L_3(2, 2, 1)$, $C_3(2, 2, 1)$ correspond to one another under the change of language which interchanges $\xrightarrow{1}$ with its reverse. So up to a change of language the complete set of constraints on triangles in Γ corresponds to entry #6 or #7 from our table.

In particular, taking Γ is itself to be 3-constrained, and in an appropriate language, it must then be isomorphic with the corresponding entry.

The proof of Proposition 21.3.2 involves direct amalgamation arguments and is broken up into a number of steps. First we deal with the triangle types $L_3(1, 2, 1)$, $L_3(1, 2, 2)$, $L_3(2, 1, 1)$, and $L_3(2, 1, 2)$, as well as L_n^1 (transitive of order n and arc type 1), and thus in particular $L_3(1, 1, 1)$.

Notation 21.3.4. We now make amalgamation arguments illustrated by diagrams. In our diagrams, 1-arcs are represented by ordinary arrows, and 2-arcs by grey arrows. (Some labels may also be shown, but not systematically.)

Lemma 21.3.5. *Suppose that Γ is a primitive non-degenerate homogeneous 2-multi-tournament in which $\xrightarrow{1}$ is transitive and there is no $\emptyset$-definable linear order. Then for all n, L_n^1 is realized in Γ, and the triangle types $L_3(1, 2, 2)$ and $L_3(2, 1, 2)$ are realized in Γ.*

PROOF. That L_n^1 is realized is straightforward: $\xrightarrow{1}$ is a partial order and the relation is non-empty, so by homogeneity there is no maximal element.

We turn to $L_3(1, 2, 2)$ and $L_3(2, 1, 2)$.

CLAIM 1. Γ *realizes at least one of the triangle types* $L_3(1, 2, 2)$ *and* $L_3(2, 1, 2)$.

These triangle types involve a vertex which via the relation $\xrightarrow{2}$ either dominates or is dominated by a 1-arc.

In what follows, we write

$$x \xrightarrow{2} y$$

to mean that there is a 2-arc between x and y: $x \xrightarrow{2} y$ or $y \xrightarrow{2} x$. Similarly, $x \xrightarrow{2} y, z$ means that $x \xrightarrow{2} y$ and $x \xrightarrow{2} z$.

Let u, v, w be a triangle in Γ of type $L_3(1, 1, 1)$. By primitivity the reflexive extension of the relation $\xrightarrow{2}$ is not an equivalence relation, so by homogeneity there is $a \in \Gamma$ with $a \xrightarrow{2} u, w$. As $\xrightarrow{1}$ is transitive we find $a \xrightarrow{2} v$. This gives us at least one of the triangle types $L_3(1, 2, 2)$ or $L_3(2, 1, 2)$, by "majority vote" among the 2-arcs.

Thus the claim holds.

By symmetry (change of language) we may suppose the triangle type $L_3(2, 1, 2)$ is realized, and we must show that $L_3(1, 2, 2)$ is realized as well.

We make amalgamation arguments.

If the triangle type $C_3(2, 2, 1)$ is realized in Γ, we use the simple amalgamation shown, with factors $C_3(2, 2, 1)$ and $L_3(2, 1, 2)$ (1-arcs: black; 2-arcs: grey).

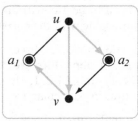

$$C_3(2, 2, 1), \; L_3(2, 1, 2)$$

Transitivity of $\xrightarrow{1}$ forces the completion to contain another 2-arc, making a copy of $L_3(1, 2, 2)$.

If $C_3(2, 2, 1)$ is not realized, then as by hypothesis $\xrightarrow{1} \cup \xrightarrow{2}$ is not linear, $C_3(2, 2, 2)$ must be realized. In this case we use the amalgamation with factors $C_3(2, 2, 2)$ and $L_3(2, 1, 2)$ shown in the following diagram.

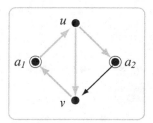

In this case as well an added 2-arc results, since $C_3(2, 2, 1)$ is forbidden, and furthermore the orientation is such as to give a copy of $L_3(1, 2, 2)$.   □

LEMMA 21.3.6. *Suppose that Γ is a primitive non-degenerate homogeneous 2-multi-tournament in which the relation $\xrightarrow{1}$ is transitive and there is no $\emptyset$-definable linear order. Then the triangle types $L_3(1, 2, 1)$ and $L_3(2, 1, 1)$ are realized in Γ.*

PROOF. Since we assume primitivity, at least one of the triangle types $L_3(1, 2, 1)$ or $L_3(2, 1, 1)$ must be realized. As our hypotheses and conclusion are preserved by reversal of the type $\xrightarrow{1}$ we may assume

$$L_3(1, 2, 1) \text{ is realized,}$$

and we aim at $L_3(2, 1, 1)$.

We may form the following amalgamation diagram with factors $L_3(1, 2, 1)$ and $L_3(1, 2, 2)$.

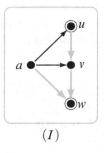

$$(I)$$

If (u, w) is a 1-arc then the orientation is $w \xrightarrow{1} u$ and the configuration $L_3(2, 1, 1)$ results. So we suppose that (u, w) forms a 2-arc in the amalgam.

If $w \xrightarrow{2} u$ then we can form the following amalgam (Figure 5), in which both factors are isomorphic to the configuration just obtained, with $\text{tp}(a_1/uvw) = \text{tp}(a/uvw)$ and $\text{tp}(a_2/vwu) = \text{tp}(a/uvw)$.

In view of the transitivity of $\xrightarrow{1}$, the points u, w force any completion of this diagram to give (a_1, a_2) the type 2 in some orientation. Then the point v produces a triangle of the desired type.

Now suppose the result of amalgamation (I) has $u \xrightarrow{2} w$. We then consider the following amalgamation, with factors $L_3(2, 1.2)$ and $L_3(1, 2, 1)$.

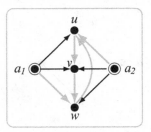

FIGURE 5. Amalgamation $(I) + (I)$.

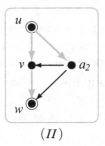

(II)

If the resulting configuration has $u \xrightarrow{1} w$ then we have a copy of $L_3(2, 1, 1)$. If it has $u \xrightarrow{2} w$ then this configuration may be combined with the result of (I) in an amalgamation diagram much like the above, apart from the orientation of some 2-arcs, and we conclude as above.

Finally, if $w \xrightarrow{2} u$ then this is similar to our first case: we may form an amalgamation diagram from two copies of this configuration, with $\mathrm{tp}(a_1/uvw) = \mathrm{tp}(a_2/vwu)$.

This completes the proof   □

Now we take a slight detour before dealing with the triangle types $L_3(2, 2, 2)$ and $C_3(2, 2, 2)$.

LEMMA 21.3.7. *Suppose that Γ is a primitive non-degenerate homogeneous 2-multi-tournament in which $\xrightarrow{1}$ transitive and there is no $\emptyset$-definable linear order, and that A is a finite configuration contained in Γ. Then there are elements a_1, a_2, a_3 in Γ satisfying the following conditions.*

$$a_1 \xrightarrow{1} A \qquad\qquad A \xrightarrow{1} a_2 \qquad\qquad a_3 \xrightarrow{2} A$$

where $\xrightarrow{2}$ signifies some mixture of $\xrightarrow{2}$ and $\xleftarrow{2}$.

PROOF. We first prove this for a_1, by induction on $|A|$, taking as the base case $|A| = 2$, in which case it is covered by Lemmas 21.3.5 and 21.3.6.

When $|A| = n > 2$ we take $B \subseteq A$ of cardinality $n - 1$ and $b \xrightarrow{1} B$ in Γ, then $a \in \Gamma$ satisfying $a \xrightarrow{1} b, A \setminus B$ to conclude.

We argue similarly for a_2.

Now we turn to a_3. With A fixed and a_1, a_2 chosen correspondingly, we may find a_3 with $a_3 \xrightarrow{2} a_1, a_2$; thus a_1, a_2, a_3 form a triangle of type $L_3(2, 1, 2)$. This then forces $a_3 \xrightarrow{2} A$ for some choice of orientations.   □

LEMMA 21.3.8. *Suppose that Γ is a primitive non-degenerate homogeneous 2-multi-tournament in which $\xrightarrow{1}$ is transitive and there is no $\emptyset$-definable linear order.*

Then the triangle type $L_3(2, 2, 2)$ is realized in Γ.

PROOF. By the previous lemma, Γ contains an infinite tournament of type $\xrightarrow{2}$. The result follows.   □

LEMMA 21.3.9. *Suppose that Γ is a primitive non-degenerate homogeneous 2-multi-tournament in which $\xrightarrow{1}$ is transitive and there is no $\emptyset$-definable linear order.*

Then $C_3(222)$ embeds into Γ.

PROOF. If one of the triangle types $L_3(2, 2, 1)$ or $C_3(2, 2, 1)$ is not realized in Γ, then since neither $\xrightarrow{1} \cup \xrightarrow{2}$ nor $\xrightarrow{1} \cup \xleftarrow{2}$ is transitive, $C_3(222)$ must embed in Γ. So we will suppose that both of the triangle types

$$L_3(2, 2, 1), \ C_3(2, 2, 1)$$

are realized in Γ.

We make an amalgamation with base

$$(u_1, u_2, u_3, u_4) \cong L_4^1$$

and with extensions by a_1, a_2 as shown.

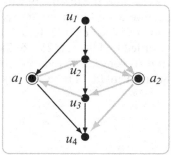

Here u_1, u_4 force the type of a_1, a_2 to be $\xrightarrow{2}$ in some orientation and u_2, u_3 force $C_3(222)$ to be realized.

So it suffices to show that the factors of this diagram embed into Γ. As $\xrightarrow{1}$ is transitive, this reduces to the various triangles containing a_1 or a_2 and two consecutive elements u_i:

$a_1 u_1 u_2$	$a_1 u_2 u_3$	$a_1 u_3 u_4$;
$a_2 u_1 u_2$	$a_2 u_2 u_3$	$a_2 u_3 u_4$.

These are of the form $L_3(121)$, $L_3(211)$, $L_3(122)$, $L_3(221)$, $L_3(212)$, or $C_3(221)$.

These 2-multi-tournaments embed in Γ either by assumption or by Lemmas 21.3.5 and 21.3.6, so the lemma follows. □

Recall now that IC_3 is the tournament consisting a point dominating an oriented 3-cycle, C_3I is its reversal, and $(IC_3)^2$ or $(C_3I)^2$ is the same tournament with the arcs interpreted as 2-arcs.

LEMMA 21.3.10. *Suppose that* Γ *is a homogeneous primitive 2-multi-tournament in which* $\xrightarrow{1}$ *is transitive and there is no* $\emptyset$-*definable linear order. Suppose also that one of the configurations* $(IC_3)^2$ *or* $(C_3I)^2$ *embeds in* Γ. *Then at least one of the triangle types* $L_3(2,2,1)$ *or* $C_3(2,2,1)$ *is realized in* Γ.

PROOF. We consider the following amalgamation diagram.

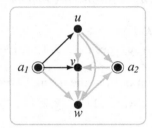

If the arc between a_1 and a_2 is a 1-arc then the vertex w forces the orientation to be $a_1 \xrightarrow{1} a_2$ and we have $L_3(2,2,1)$, so we conclude. If the arc between a_1 and a_2 is a 2-arc then u or v gives $L_3(2,2,1)$ or $C_3(2,2,1)$, and we again conclude. So it suffices to show that the factors of this diagram are realized.

In the 3-constrained case we would be done at this point, as the triangles involved in the diagrams are of types $L_3(1,2,1)$, $L_3(1,2,2)$, $L_3(2,2,2)$, and $C_3(2,2,2)$ and these are afforded by Lemmas 21.3.5, 21.3.6, 21.3.8, and 21.3.9.

To reach a contradiction in general, it remains to construct the factors of this amalgamation diagram in Γ. These factors are shown separately below.

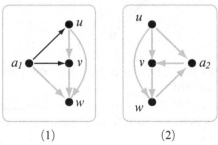

(1)　　　　　　(2)

The factor omitting a_2 shown on the left may be considered as an amalgamation diagram with the type of (u, w) to be determined. The only solution compatible with our assumptions is the one shown: $u \xrightarrow{2} w$. As this diagram has subfactors $L_3(1, 2, 1)$ and $L_2(1, 2, 2)$ the amalgam must exist, so this factor must exist.

The factor omitting a_1, shown on the right, has the form $u \xrightarrow{2} C_3^2$. If $(IC_3)^2$ is realized in Γ then the proof is complete, and if $(C_3I)^2$ is realized in Γ then we obtain the same result by reversing the orientations of the two 2-types. $\square$

PROOF OF PROPOSITION 21.3.2. We assume $\xrightarrow{1}$ gives a partial order on the primitive non-degenerate homogeneous 2-multi-tournament Γ, and there is no $\emptyset$-definable linear order on Γ.

Then Lemmas 21.3.5, 21.3.6, 21.3.8, and 21.3.9, give respectively $L_3(1, 1, 1)$, $L_3(1, 2, 2)$, $L_3(2, 1, 2)$; $L_3(1, 2, 1)$, $L_3(2, 1, 1)$, $L_3(2, 2, 2)$, and $C_3(222)$. If $(IC_3)^2$ or $(C_3I)^2$ is realized in Γ, then Lemma 21.3.10 gives $L_3(2, 2, 1)$ or $C_3(2, 2, 1)$ as well.

Under the hypothesis of 3-constraint, both $(IC_3)^2$ and $(C_3)^2$ are realized in Γ. $\square$

This analysis continues in greater generality in §22.2.

21.4. Completeness: no partial orders

We turn to the more complex, and less intuitive, case in which there are no nontrivial $\emptyset$-definable partial orders: that is, $\xrightarrow{1}$ and $\xrightarrow{2}$ are not transitive, and no union of 2-types is a linear order. In this case Proposition 21.3.1 has three parts, which we now state separately.

PROPOSITION 21.4.1 (NPO$_{3c}$: $\widetilde{\mathbb{S}(3)}$). *Suppose that Γ is an infinite primitive non-degenerate homogeneous 3-constrained 2-multi-tournament with no nontrivial $\emptyset$-definable partial order. Suppose that neither of the triangle types*

$$C_3(1, 1, 1), C_3(2, 2, 2)$$

is realized in Γ.

Then up to a change of language Γ satisfies the same triangle constraints as $\widetilde{\mathbb{S}(3)}$.

Of course, in this case we expect Γ to be isomorphic with $\widetilde{\mathbb{S}(3)}$ after identifying the languages appropriately. This appears to be a particularly tractable case for complete analysis, but we will not follow up on such issues here, as they lie just outside the scope of our discussion.

PROPOSITION 21.4.2 (NPO$_{3c}$: #8, #10, $\mathbb{S}(4)$). *Suppose that Γ is an infinite primitive non-degenerate homogeneous 3-constrained 2-multi-tournament with no nontrivial $\emptyset$-definable partial order, not associated with a free amalgamation class.*

Suppose that the triangle type

$$C_3(1, 1, 1)$$

is forbidden and the triangle type

$$C_3(2, 2, 2)$$

is realized. Then up to a change of language Γ is determined by the constraints given in one of entries #8 or #10 of our table, or Γ is $\mathbb{S}(4)$ (#12).

PROPOSITION 21.4.3 (NPO$_{3c}$: #9). *Suppose that Γ is an infinite primitive nondegenerate homogeneous 3-constrained 2-multi-tournament in with no nontrivial $\emptyset$-definable partial order, and which does not have free amalgamation. Suppose that the triangle types*

$$C_3(1, 1, 1), \; C_3(2, 2, 2)$$

are both realized in Γ. Then up to a change of language, Γ is determined by the constraints in entry #9 of our table.

We give this as Table 21.2 below, in terms of the triangle constraints involved.

I	$C_3(111)$, $C_3(222)$ forbidden
#11, $\widetilde{\mathbb{S}(3)}$	$L_3(111)$, $L_3(122)$, $L_3(212)$ forbidden
II	$C_3(111)$ forbidden, $C_3(222)$ realized
#8	$C_3(112)$ forbidden
#10	$C_3(112)$, $L_3(111)$ forbidden
#12, $\mathbb{S}(4)$	$C_3(112)$, $L_3(121)$, $L_3(211)$, $L_3(221)$, $L_3(222)$ forbidden
III	$C_3(111)$, $C_3(222)$ realized
#9	$C_3(112)$, $L_3(111)$ forbidden

TABLE 21.2. Primitive 3-constrained 2-multi-tournaments with no nontrivial $\emptyset$-definable partial order, and without free amalgamation.

21.4.1. Proof of Proposition 21.4.1. We analyze infinite primitive nondegenerate homogeneous 2-multi-tournaments with no nontrivial $\emptyset$-definable partial order and with triangle types $C_3(1, 1, 1)$ and $C_3(2, 2, 2)$ forbidden, assuming 3-constraint when needed (or convenient).

Table 21.3 below summarizes the steps involved (Lemmas 21.4.4–21.4.9).

LEMMA 21.4.4. *Suppose that Γ is a homogeneous primitive 2-multi-tournament in which there is no nontrivial $\emptyset$-definable partial order and triangles of types*

$$C_3(1, 1, 1), \; C_3(1, 1, 2)$$

	Lemma 21.3.9: *Target*	

Forbid $L_3(111)$, $L_3(122)$, $L_3(212)$, $C_3(111)$, $C_3(222)$.

Realize $L_3(112)$, $L_3(121)$, $L_3(211)$, $L_3(221)$, $L_3(222)$, $C_3(112)$, $C_3(221)$.

$\overset{1}{\longrightarrow}$ is the de-symmetrized type.

Lemma	Hypotheses	Conclusion	page
21.4.4.1	$\neg C_3(111)$, $\neg C_3(112)$	$L_3(112)$, $C_3(221)$	126
21.4.4.2	$\neg C_3(111)$, $\neg C_3(112)$, $\neg C_3(222)$	$L_3(111)$, $L_3(121)$, $L_3(122)$, $L_3(211)$, $L_3(212)$	126
21.4.5	$\neg L_3(122)$ or $\neg L_3(212)$	$L_3(222)$	130
21.4.6	$\neg C_3(111)$, $\neg C_3(222)$, 3-constraint	$C_3(112)$, $C_3(221)$, $L_3(112)$, $L_3(221)$	130
21.4.7	$\neg C_3(111)$, $\neg C_3(222)$, $C_3(221)$, $C_3(112)$	(4 Cases: a, a', b, c)	131
21.4.8	", 3-constraint	Not (a')	132
21.4.9	$\neg C_3(111)$, $\neg C_3(222)$, 3-constraint	w.l.o.g. $\neg L_3(122)$, $\neg L_3(212)$, $L_3(121)$, $L_3(211)$	133
Proposition			
21.4.1	"	$\widetilde{\mathbb{S}(3)}$-like	134

TABLE 21.3. Proof of Proposition 21.4.1.

are forbidden. Then the following hold.

1. *The triangle types*

$$L_3(1,1,2), C_3(2,2,1)$$

   are realized in Γ.
2. *If the triangle type* $C_3(2,2,2)$ *is also forbidden, then the following triangle types are realized.*

$$L_3(111), \qquad L_3(121), \qquad L_3(122), \qquad L_3(211), \qquad L_3(212).$$

PROOF. *Ad* 1. Here we assume only that the triangle types $C_3(1,1,1)$ and $C_3(1,1,2)$ are forbidden.

As $\overset{1}{\longrightarrow}$ is not transitive, $L_3(1,1,2)$ must be realized.

Suppose $C_3(2,2,1)$ is forbidden. As $\overset{1}{\longrightarrow} \cup \overset{2}{\longrightarrow}$ is not a linear order, $C_3(2,2,2)$ must embed in Γ. Then conclude via the following amalgamation.

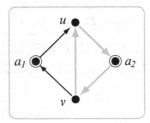

Ad 2. Now we assume in addition that $C_3(2,2,2)$ is forbidden.

CLAIM 1. $L_3(1,2,2)$ *embeds into* Γ.

We use one of the following amalgamation diagrams.

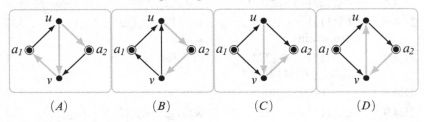

$\quad\quad\quad (A) \quad\quad\quad\quad (B) \quad\quad\quad\quad (C) \quad\quad\quad\quad (D)$

The factors are, respectively

$\quad\quad (A)\quad C_3(2,2,1),\, L_3(2,1,2);\quad\quad (B)\quad L_3(1,1,1),\, C_3(2,2,1);$
$\quad\quad (C)\quad L_3(1,2,1),\, L_3(2,2,1);\quad\quad (D)\quad L_3(1,2,1),\, C_3(2,2,1).$

If $L_3(2,1,2)$ or $L_3(1,1,1)$ embeds in Γ, we may use diagram (A) or (B) respectively, which forces either $a_1 \xrightarrow{2} a_2$ or $a_2 \xrightarrow{2} a_1$, and thus $L_3(1,2,2)$.

So suppose now that $L_3(2,1,2)$ and $L_3(1,1,1)$ are forbidden.

Considering two realizations of the type $a \xrightarrow{1} x$ shows that $L_3(1,2,1)$ is realized.

Then diagram (D) can be constructed, and under our current assumptions it forces $L_3(2,2,1)$.

Then under our current assumptions diagram (C) forces $L_3(1,2,2)$.

This proves the claim.

Our hypotheses are conserved under reversal of all arcs, and hence by symmetry, we conclude that the triangle type

$$L_3(212)$$

is also realized.

For reference we give a table of the constraints known or assumed at this point, and those still to be dealt with.

Forbidden:	$C_3(1,1,1)$, $C_3(1,1,2)$, $C_3(2,2,2)$
Realized:	$C_3(2,2,1)$, $L_3(1,1,2)$, $L_3(1,2,2)$, $L_3(2,1,2)$
Targets:	$L_3(1,1,1)$, $L_3(1,2,1)$, $L_3(2,1,1)$

CLAIM 2. $L_3(1,1,1)$ *embeds into* Γ.

We use the following amalgamation.

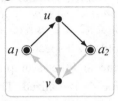

CLAIM 3. $L_3(1,2,1)$ *embeds into* Γ.

We use the following amalgamations.

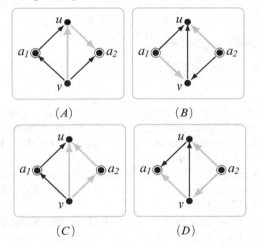

(A) (B)

(C) (D)

These have the following factors.

(A) $L_3(1,1,2), L_3(2,2,1)$. (B) $L_3(2,1,1), L_3(1,1,2)$.

(C) $L_3(1,1,2), L_3(2,2,2)$. (D) $L_3(1,1,2), L_3(2,1,2)$.

If $L_3(2,2,1)$ embeds into Γ then diagram (A) forces $L_3(1,2,1)$. So suppose $L_3(2,2,1)$ is forbidden.

If $L_3(2,1,1)$ embeds into Γ then diagram (B) forces $L_3(1,2,1)$ or $L_3(2,2,1)$, and as we assume the latter is forbidden this gives $L_3(1,2,1)$. So suppose $L_3(2,1,1)$ is also forbidden.

If $L_3(2,2,2)$ embeds into Γ then diagram (C) forces $L_3(1,2,1)$, $L_3(2,1,1)$, or $L_3(2,2,1)$, and again we conclude.

Diagram (D) has factors known to embed in Γ, and forces either $a_2 \overset{1}{\to} a_1$ or $a_2 \overset{2}{\to} a_1$, resulting in either $L_3(2, 2, 1)$ or $L_3(2, 2, 2)$. So one of the previous cases applies and the claim follows.

Finally, by symmetry, $L_3(2, 1, 1)$ also embeds into Γ.

This completes the treatment of all of the stated triangle types.   □

LEMMA 21.4.5. *Let Γ be a primitive non-degenerate homogeneous 2-multi-tournament in which one of the triangle types $L_3(1, 2, 2)$ or $L_3(2, 1, 2)$ is not realized. Then the triangle type $L_3(2, 2, 2)$ is realized.*

PROOF. If Γ is finite this holds by inspection. If Γ is infinite and primitive then for $a \in \Gamma$, a^2 is also infinite, and the assumption implies that the type $\overset{1}{\to}$ is not realized there, so the type $\overset{2}{\to}$ is realized. This gives the result.   □

LEMMA 21.4.6. *Suppose that Γ is a homogeneous primitive 3-constrained 2-multi-tournament in which there is no nontrivial $\emptyset$-definable partial order, and triangles of types*

$$C_3(1, 1, 1), C_3(2, 2, 2)$$

are forbidden. Then triangles of types

$$L_3(1, 1, 2), \qquad L_3(2, 2, 1), \qquad C_3(1, 1, 2), \qquad C_3(2, 2, 1),$$

embed in Γ.

PROOF. Our hypotheses are conserved under switching labels or reversing the orientation of a 2-type, and this group of permutations of the language acts transitively on the four triangle types listed. So it suffices to show that the triangle type $C_3(1, 1, 2)$ is realized.

If it is not realized then Lemma 21.4.4 applies and gives realizations of the triangle types $L_3(1, 1, 1)$, $L_3(1, 1, 2)$, $L_3(1, 2, 1)$, $L_3(1, 2, 2)$, $L_3(2, 1, 1)$, $L_3(2, 1, 2)$, and $C_3(2, 2, 1)$.

In the following amalgamation diagram the triangles occurring in the factors are of types $L_3(1, 1, 2)$, $L_3(1, 2, 2)$, $L_3(2, 1, 2)$, $C_3(2, 2, 1)$ and by 3-constraint these factors occur in Γ.

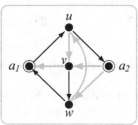

Our constraints force $a_2 \overset{2}{\to} a_1$ in the completion and then $a_1 u a_2$ has type $C_3(1, 1, 2)$.   □

Lemma 21.4.7. *Suppose that* Γ *is a homogeneous primitive 2-multi-tournament in which there is no nontrivial $\emptyset$-definable partial order and triangles of types*

$$C_3(1, 1, 1), C_3(2, 2, 2)$$

are forbidden, while triangle types

$$C_3(1, 1, 2), C_3(2, 2, 1)$$

are realized.

Then up to a permutation of the language one of the following cases applies.

(a)

Forbidden		Realized	
$L_3(1, 2, 2)$	$L_3(2, 1, 2)$	$L_3(1, 2, 1)$	$L_3(2, 1, 1)$

(a')

Forbidden		Realized	
$L_3(1, 2, 1)$	$L_3(2, 1, 2)$	$L_3(1, 2, 2)$	$L_3(2, 1, 1)$

(b)

Forbidden	Realized		
$L_3(1, 2, 1)$	$L_3(1, 2, 2)$	$L_3(2, 1, 1)$	$L_3(2, 1, 2)$

(c)

Realized			
$L_3(1, 2, 1)$	$L_3(1, 2, 2)$	$L_3(2, 1, 1)$	$L_3(2, 1, 2)$

Proof. We are considering which of the following four triangle types are realized in Γ.

$$L_3(1, 2, 1), L_3(1, 2, 2), L_3(2, 1, 1), L_3(2, 1, 2).$$

We consider the group of permutations of the language generated by switching the two labels $1, 2$ and by reversing orientations of both 2-types. This group acts transitively on this set of four triangle types, and preserves the set of assumed constraints.

Claim 1. *At least one triangle type out of each of the following pairs of triangle types is realized in* Γ.

$$L_3(1, 2, 1), L_3(1, 2, 2)$$
$$L_3(2, 1, 1), L_3(2, 1, 2).$$

We may suppose that $L_3(1, 1, 1)$ or $L_3(2, 2, 2)$ is realized, since otherwise Γ would be finite, and one can check the classification in that case. Supposing for example that $L_3(1, 1, 1)$ is realized, the following amalgamation diagram gives one of the required triangle types.

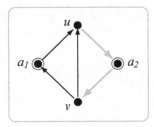

By the symmetry noted above, which also applies to the statement of the claim, we may suppose that the triangle type $L_3(2, 1, 2)$ is realized, in which case we claim that one of the triangle types $L_3(1, 2, 1)$ or $L_3(1, 2, 2)$ is realized. For this we use the following amalgamation diagram.

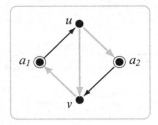

The claim follows.

Now up to a symmetry of the language we arrive at one of the cases listed, depending on whether we realize two, three, or all four of the specified triangle types.   □

LEMMA 21.4.8. *Suppose that Γ is a 3-constrained homogeneous primitive 2-multi-tournament in which there is no nontrivial $\emptyset$-definable partial order. Suppose that triangles of types*

$$C_3(1, 1, 1), C_3(2, 2, 2)$$

are forbidden, while triangle types

$$C_3(1, 1, 2), C_3(2, 2, 1)$$

are realized.
    If triangle type $L_3(2, 1, 2)$ is forbidden, then triangle type $L_3(1, 2, 1)$ is realized.

PROOF. By Lemma 21.4.7 the case that needs to be eliminated has the following constraints (case (a')).

$$L_3(1, 2, 1), L_3(2, 1, 2) \text{ are forbidden.}$$
$$L_3(1, 2, 2), L_3(2, 1, 1) \text{ are realized.}$$

By considering the configuration aub with $a \xrightarrow{1} u$ and $b \xrightarrow{2} u$, our constraints force $b \xrightarrow{1} a$ or $a \xrightarrow{2} b$ and thus one of the types $L_3(1, 1, 2)$ or $L_3(2, 2, 1)$ is realized. By symmetry we may suppose that $L_3(1, 1, 2)$ is realized. Then we use the following amalgamation diagram.

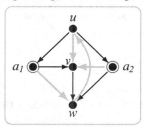

The triangles occurring in the factors are of types $L_3(1, 1, 2)$, $L_3(1, 2, 2)$, $L_3(2, 1, 1)$, $C_3(1, 1, 2)$, and $C_3(2, 2, 1)$. By 3-constraint, these factors occur in Γ. In any completion of the diagram, one of the vertices u, v, w completes a triangle of type $L_3(1, 2, 1)$.   □

LEMMA 21.4.9. *Suppose that Γ is a homogeneous primitive 3-constrained 2-multi-tournament in which there is no nontrivial $\emptyset$-definable partial order and triangles of types*

$$C_3(1, 1, 1), \ C_3(2, 2, 2)$$

are forbidden.

Then up to a permutation of the language we have

The triangle types $L_3(1, 2, 2)$ and $L_3(2, 1, 2)$ are forbidden.

The triangle types $L_3(1, 2, 1)$ and $L_3(2, 1, 1)$ are realized.

PROOF. By Lemma 21.4.6 the following triangle types are realized.

$$L_2(1, 1, 2) \qquad L_3(2, 2, 1) \qquad C_3(1, 1, 2) \qquad C_3(2, 2, 1).$$

So Lemma 21.4.7 applies and provides four possibilities, labeled as (a, a', b, c). The case indicated in the statement of the lemma is case (a).

With the assumption of 3-constraint, Lemma 21.4.8 We must also eliminate cases (b, c), again assuming 3-constraint. This is the subject of the next two claims.

CLAIM 1. *Under the stated assumptions we cannot have*

$L_3(1, 2, 1)$ is forbidden.

$L_3(1, 2, 2)$, $L_3(2, 1, 1)$, $L_3(2, 1, 2)$ are realized.

We may suppose that Γ is infinite as otherwise the claim follows by inspection. In particular for any vertex a we have at least two vertices u, v in a^1. By assumption the arc between u, v is a 1-arc and therefore (a, u, v) has type $L_3(1, 1, 1)$. So this triangle type must be realized.

We consider the following amalgamation diagram.

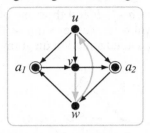

The triangle types involved are $L_3(1,1,1)$, $L_3(1,1,2)$, $L_3(2,1,1)$, $C_3(1,1,2)$, and $C_3(2,2,1)$. By the hypothesis of 3-constraint the factors of this diagram occur in Γ and therefore there is a completion in Γ.

If the arc between a_1 and a_2 is a 1-arc then vertex v or w provides a copy of $C_3(1,1,1)$. If it is a 2-arc then u provides $L_3(1,2,1)$. So we have a contradiction.

CLAIM 2. *Under the stated assumptions we cannot have all of*

$$L_3(1,2,1), L_3(1,2,2), L_3(2,1,1), L_3(2,1,2),$$

realized.

Note that in view of the remarks above, this would mean that all triangles which are not monochromatic are realized.

The argument involves our most elaborate amalgamation diagram so far, with 6 vertices a, b, u_1, u_2, u_3, u_4, as follows. The structure on u_1, u_2, u_3, u_4 is shown without orientations since they will not matter.

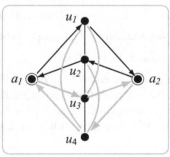

Since there are no monochromatic triangles in this diagram, both factors occur in Γ, by 3-constraint.

In a completion, the 4-cycles containing $u_1 u_2$ or u_3, u_4 would force either $C_3(1,1,1)$ or $C_3(2,2,2)$ to occur, for a contradiction.   □

PROOF OF PROPOSITION 21.4.1. We suppose that triangle types $C_3(1,1,1)$ and $C_3(2,2,2)$ are forbidden, and we must show that the language can be chosen so that the triangle constraints satisfied by Γ are those of $\widetilde{\mathbb{S}}(3)$.

Lemma 21.4.6 gives the realization of the other two oriented 3-cycles, as well as $L_3(1, 1, 2)$ and $L_3(2, 2, 1)$. Then by Lemma 21.4.9, up to a choice of language we may suppose that $L_3(1, 2, 2)$, $L_3(2, 1, 2)$ are forbidden and $L_3(1, 2, 1)$, $L_3(2, 1, 1)$ are realized.

By Lemma 21.4.5 the triangle type $L_3(2, 2, 2)$ is also realized.

CLAIM 1. *The triangle type $L_3(1, 1, 1)$ is not realized.*

We consider the following amalgamation diagram.

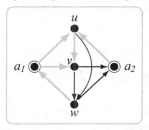

This involves the triangle types $L_3(1, 1, 1)$, $L_3(2, 1, 1)$, $L_3(2, 2, 2)$, $C_3(1, 1, 2)$, $C_3(2, 2, 1)$. As these are present in Γ, the amalgamation diagram has a completion in Γ. But the point u forces a_1, a_2 to be joined by a 2-arc and the points v, w prevent this.

This proves the claim.

This completes the analysis of all triangle types.   □

21.4.2. Proposition 21.4.2.

PROPOSITION 21.4.2 (NPO$_{3c}$: #8, #10, $\mathbb{S}(4)$). *Suppose that Γ is an infinite primitive non-degenerate homogeneous 3-constrained 2-multi-tournament with no nontrivial $\emptyset$-definable partial order, not associated with a free amalgamation class.*

Suppose that the triangle type

$$C_3(1, 1, 1)$$

is forbidden and the triangle type

$$C_3(2, 2, 2)$$

is realized. Then up to a change of language Γ is determined by the constraints given in one of entries #8 or #10 of our table, or Γ is isomorphic to $\mathbb{S}(4)$ (#12).

We repeat the relevant lines of the table.

#8	$C_3(111)$, $C_3(112)$ forbidden.
#10	$C_3(111)$, $C_3(112)$, $L_3(111)$ forbidden.
#12, $\mathbb{S}(4)$	$C_3(111)$, $C_3(112)$, $L_3(121)$, $L_3(211)$, $L_3(221)$, $L_3(222)$ forbidden.

We organize the analysis as follows. When $L_3(2, 2, 1)$ is forbidden we identify $\mathbb{S}(4)$ (Proposition 21.4.10). When $L_3(1, 1, 1)$ is forbidden we identify entry

#10 (Proposition 21.4.19). When $L_3(1, 1, 1)$ and $L_3(2, 2, 1)$ are realized we identify entry #8 (Proposition 21.4.21).

Of course, when we assume 3-constraint, identifying the structure reduces to identifying the triangles which embed in the structure. In fact, in Proposition 21.4.19 we work more generally: without assuming 3-constraint, we determine the pattern of forbidden triangles.

21.4.3. The case of $\mathbb{S}(4)$. Our target in this subsection is the following.

PROPOSITION 21.4.10. *Suppose that Γ is a 3-constrained homogeneous primitive 2-multi-tournament for which the triangle types*

$$C_3(1, 1, 1), \; L_3(2, 2, 1)$$

are forbidden and the triangle type $C_3(2, 2, 2)$ is realized.
Then Γ is isomorphic to $\mathbb{S}(4)$.

LEMMA 21.4.11. *Suppose that Γ is a homogeneous primitive 2-multi-tournament. Suppose that the triangle types*

$$C_3(1, 1, 1), \; L_3(2, 2, 1)$$

are forbidden and the triangle type $C_3(2, 2, 2)$ is realized.
Then the triangle types $L_3(1, 2, 2)$ and $L_3(2, 1, 2)$ are realized.

PROOF. As the hypotheses are preserved by reversal of the orientation of both 2-types, it suffices to treat the case of $L_3(1, 2, 2)$.

We suppose the contrary:

Triangle types $L_3(1, 2, 2)$, $L_3(2, 2, 1)$, and $C_3(1, 1, 1)$ are forbidden.

Triangle type $C_3(2, 2, 2)$ is realized.

CLAIM 1. *Triangle type $C_3(2, 2, 1)$ is forbidden.*

Otherwise we make the following amalgamation.

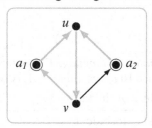

This has no completion consistent with the constraints, so the claim follows.

As Γ is primitive and triangle types $L_3(1, 2, 2)$, $L_3(2, 2, 1)$, and $C_3(2, 2, 1)$ are forbidden, the triangle type $L_3(2, 1, 2)$ must be realized. Then we can form the following amalgamation diagram.

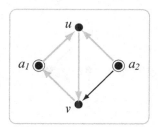

But this has no completion consistent with the constraints.
This completes the proof.   □

LEMMA 21.4.12. *Suppose that* Γ *is a homogeneous primitive 2-multi-tournament with no nontrivial* $\emptyset$-*definable partial order. Suppose that the triangle types* $C_3(1,1,1)$ *and* $L_3(2,2,1)$ *are forbidden and the triangle type* $C_3(2,2,2)$ *is realized. Then the triangle type* $L_3(1,1,2)$ *is realized.*

PROOF. Suppose the contrary. As the 2-type $\overset{1}{\longrightarrow}$ is not transitive, the triangle type $C_3(1,1,2)$ must be realized. Consider the following amalgamation diagram.

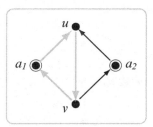

As this has no completion consistent with the constraints the lemma follows.
                                                                      □

LEMMA 21.4.13. *Suppose that* Γ *is a 3-constrained homogeneous primitive 2-multi-tournament with no nontrivial* $\emptyset$-*definable partial order. Suppose that the triangle types* $C_3(1,1,1)$ *and* $L_3(2,2,1)$ *are forbidden and the triangle type* $C_3(2,2,2)$ *is realized. Then the triangle type* $L_3(1,1,1)$ *is realized.*

PROOF. Assume the contrary. For any vertex a, if we consider two vertices dominated by a, or dominating a, with respect to 1-arcs, then we see that the triangle types

$$L_3(1,2,1),\ L_3(2,1,1)$$

must be realized.
Consider the following amalgamation diagram.

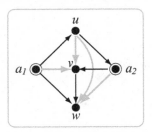

The triangles occurring in the factors are of the forms $L_3(1,1,2)$, $L_3(1,2,1)$, $L_3(1,2,2)$, $L_3(2,1,1)$, and $L_3(2,1,2)$.

In view of Lemmas 21.4.11 and 21.4.12 and the hypothesis of 3-constraint, this diagram must have a completion in Γ. Vertices v, w force a_1, a_2 to be related by a 1-arc and then $a_1 u a_2$ has the form $L_3(1,1,1)$.

This proves the lemma.   □

LEMMA 21.4.14. *Suppose that Γ is a 3-constrained homogeneous primitive 2-multi-tournament with no nontrivial $\emptyset$-definable partial order. Suppose that the triangle types $C_3(1,1,1)$ and $L_3(2,2,1)$ are forbidden and the triangle type $C_3(2,2,2)$ is realized. Then the triangle type $C_3(2,2,1)$ is realized.*

PROOF. Consider the following amalgamation diagram.

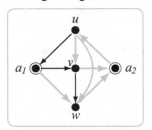

The triangles occurring in the factors are of types $L_3(1,1,2)$, $L_3(1,2,2)$, $L_3(2,1,2)$, and $C_3(2,2,2)$, which are realized in Γ by Lemmas 21.4.11 and 21.4.12. So by 3-constraint this diagram has a completion in Γ. Any completion would involve $L_3(2,2,1)$ or $C_3(2,2,1)$, and the former is forbidden. The lemma follows.   □

We take stock of the analysis so far.

LEMMA 21.4.15. *Suppose that Γ is a 3-constrained homogeneous primitive 2-multi-tournament with no nontrivial $\emptyset$-definable partial order. Suppose that the triangle types $C_3(1,1,1)$ and $L_3(2,2,1)$ are forbidden and the triangle type $C_3(2,2,2)$ is realized. Then $\mathbb{S}(4)$ embeds into Γ.*

PROOF. The triangle types embedding in $\mathbb{S}(4)$ are

$$L_3(1,1,1), L_3(1,1,2), L_3(1,2,2);$$
$$L_3(2,1,2), C_3(2,2,1), C_3(2,2,2).$$

These are afforded by the hypothesis and by Lemmas 21.4.11, 21.4.12, 21.4.13, and 21.4.14.   □

LEMMA 21.4.16. *Suppose that* Γ *is a 3-constrained homogeneous 2-multi-tournament for which the triangle types* $C_3(1, 1, 1)$ *and* $L_3(2, 2, 1)$ *are forbidden and* $\mathbb{S}(4)$ *embeds into* Γ. *Then at least one of* $L_3(1, 2, 1)$ *or* $C_3(1, 1, 2)$ *is not realized in* Γ.

PROOF. Consider the following amalgamation diagram.

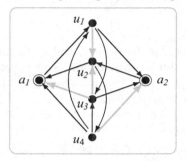

The triangle types involved are $L_3(1, 1, 1)$, $L_3(1, 1, 2)$, $L_3(1, 2, 1)$, $L_3(1, 2, 2)$, $L_3(2, 1, 2)$, and $C_3(1, 1, 2)$.

As no completion satisfies the constraints and triangles of types $L_3(1, 1, 1)$, $L_3(1, 1, 2)$, $L_3(1, 2, 2)$, $L_3(2, 1, 2)$ are realized in Γ, the lemma follows.   □

LEMMA 21.4.17. *Suppose that* Γ *is a 3-constrained homogeneous 2-multi-tournament for which the triangle types* $C_3(1, 1, 1)$ *and* $L_3(2, 2, 1)$ *are forbidden and* $\mathbb{S}(4)$ *embeds into* Γ.

Then the triangle type $C_3(1, 1, 2)$ *is not realized in* Γ.

PROOF. Supposing the contrary, the triangle type $L_3(1, 2, 1)$ is forbidden by Lemma 21.4.16.

CLAIM 1. *The triangle type* $L_3(2, 1, 1)$ *is forbidden.*

Consider the following amalgamation diagram.

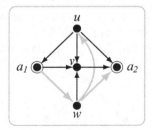

The triangle types involved are $L_3(1, 1, 1)$, $L_3(1, 1, 2)$, $L_3(2, 1, 1)$, $L_3(2, 1, 2)$, and $L_3(2, 2, 1)$. All of these with the exception of $L_3(2, 1, 1)$ embed into Γ. But a completion in Γ would produce a triangle (u, a_1, a_2) of type $(1, 2, 1)$.

This proves the claim.

Now consider the following amalgamation diagram.

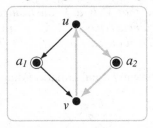

The completion has a 1-arc between a_1 and a_2, forming a triangle of type $L_3(1, 2, 1)$ or $L_3(2, 1, 1)$, for a contradiction. This proves the lemma.   □

LEMMA 21.4.18. *Suppose that Γ is a 3-constrained homogeneous 2-multi-tournament for which the triangle types $C_3(1, 1, 1)$ and $L_3(2, 2, 1)$ are forbidden and $\mathbb{S}(4)$ embeds into Γ. Then the triangle types $L_3(1, 2, 1)$, $L_3(2, 1, 1)$ and $L_3(2, 2, 2)$ are not realized in Γ.*

PROOF. We know that the triangle type $C_3(1, 1, 2)$ is forbidden by Lemma 21.4.17.

CLAIM 1. *The triangle types $L_3(1, 2, 1)$ and $L_3(2, 1, 1)$ are forbidden.*

Consider the following amalgamation diagram.

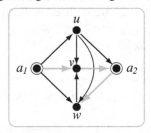

This involves triangle types $L_3(1, 1, 1)$, $L_3(1, 1, 2)$, $L_3(1, 2, 1)$, and $C_3(2, 2, 1)$ and has no completion in Γ.

Thus the triangle type $L_3(1, 2, 1)$ is forbidden. By symmetry, reversing both 2-types, the triangle type $L_3(2, 1, 1)$ is forbidden.

Finally, consider the following amalgamation diagram.

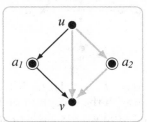

This has no completion in Γ and so $L_3(2,2,2)$ is forbidden. $\square$

PROOF OF PROPOSITION 21.4.10. By Lemma 21.4.15 every triangle embedding in $\mathbb{S}(4)$ embeds into Γ.

By Lemmas 21.4.17 and 21.4.18 no other triangles embed into Γ. The result follows. $\square$

21.4.4. Entry #10. In this subsection we aim at the following.

PROPOSITION 21.4.19. *Suppose that Γ is an infinite primitive homogeneous 2-multi-tournament for which the triangle types*

$$C_3(1,1,1),\ C_3(1,1,2),\ L_3(1,1,1),$$

are forbidden, and there is no $\emptyset$-definable linear order.
Then all other triangle types are realized, as in entry #10 of our table.

Here there is a symmetry: our constraint set is preserved by reversing both 2-types.

LEMMA 21.4.20. *Suppose that Γ is an infinite homogeneous primitive 2-multi-tournament for which the triangle types*

$$C_3(1,1,1),\ C_3(1,1,2),\ L_3(1,1,1)$$

are forbidden.
Then the triangle types

$$L_3(1,1,2),\ L_3(1,2,1),\ L_3(1,2,2),\ L_3(2,1,1),\ L_3(2,1,2),\ L_3(2,2,2)$$

are realized.

PROOF. The unique completion of the configuration $a \xrightarrow{1} u \xrightarrow{1} b$ is $L_3(1,1,2)$.

Fixing a point a in Γ and considering a^1, $a^{1^{op}}$ yields triangle types $L_3(1,2,1)$ and $L_3(2,1,1)$. (Notation as in Notation 20.1.1, with respect to $\xrightarrow{1}$ and $\xleftarrow{1}$.)

Since $L_3(1,1,1)$ is forbidden, $L_3(2,2,2)$ is realized.

It remains to consider triangle types $L_3(1,2,2)$ and $L_3(2,1,2)$.

The following amalgamation diagram gives triangle type $L_3(1,2,2)$.

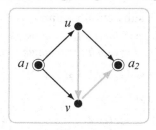

Since our constraints are preserved by reversing the orientations of both 2-types, we also have triangle type $L_3(2,1,2)$. $\square$

PROOF OF PROPOSITION 21.4.19. By Lemma 21.4.20, triangle types

$$L_3(1,1,2), \quad L_3(1,2,1), \quad L_3(1,2,2), \quad L_3(2,1,1), \quad L_3(2,1,2),$$
$$L_3(2,2,2),$$

are realized, leaving the triangle types

$$L_3(2,2,1), C_3(2,2,1), C_3(2,2,2),$$

still to be dealt with.

CLAIM 1. *Type $C_3(2,2,1)$ is realized.*

Suppose $C_3(2,2,1)$ is forbidden. Since there is no $\emptyset$-definable linear order, the triangle type $C_3(2,2,2)$ is realized.

But then it suffices to complete the diagram below to get triangle type $C_3(2,2,1)$.

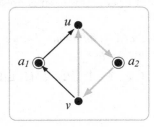

The claim follows.

Now the following diagrams give $L_3(2,2,1)$ and $C_3(2,2,2)$.

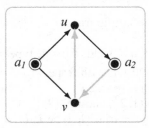 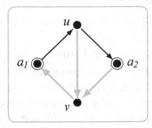

21.4.5. Entry #8. In this subsection we aim at the following.

PROPOSITION 21.4.21. *Suppose that Γ is an infinite 3-constrained homogeneous primitive 2-multi-tournament for which the triangle types*

$$C_3(1,1,1), C_3(1,1,2)$$

are forbidden, the triangle types

$$L_3(1,1,1), L_3(2,2,1), C_3(2,2,1)$$

are realized, and there is no nontrivial $\emptyset$-definable partial order.

Then all other triangle types are realized, as in entry #8 of our table.

LEMMA 21.4.22. *Suppose that* Γ *is an infinite homogeneous primitive 2-multi-tournament for which the triangle types*

$$C_3(1,1,1), \quad C_3(1,1,2)$$

are forbidden, the triangle types

$$L_3(1,1,1), \quad L_3(2,2,1), \quad C_3(2,2,1)$$

are realized, and there is no nontrivial $\emptyset$-definable partial order.
Then the triangle types

$$L_3(1,1,2), \quad L_3(1,2,1), \quad L_3(2,1,1)$$

are realized.

PROOF. As the 2-type $\xrightarrow{1}$ is not transitive, the type $L_3(1,1,2)$ must be realized.

The following amalgamation diagram gives $L_3(1,2,1)$.

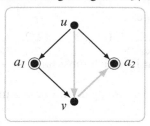

As our constraints are preserved by reversal of both 2-types, the type $L_3(2,1,1)$ is also realized. $\square$

PROOF OF PROPOSITION 21.4.21. By Lemma 21.4.22 and the hypothesis the following triangle types are realized.

$$L_3(1,1,1), \quad L_3(1,1,2), \quad L_3(1,2,1), \quad L_3(2,1,1), \quad L_3(2,2,1).$$
$$C_3(2,2,1).$$

This leaves the following types to be considered.

$$L_3(1,2,2), \quad L_3(2,1,2), \quad L_3(2,2,2), \quad C_3(2,2,2).$$

Consider the following amalgamation diagram.

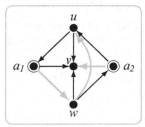

The triangle types involved are
$$L_3(1,1,1),\ L_3(1,1,2),\ L_3(1,2,1),\ \text{and}\ L_3(2,1,1),$$
which are realized by Γ.

The only possible completion has $a_2 \xrightarrow{2} a_1$ and thus $L_3(2,1,2)$ is realized. As the constraints are closed under reversal, also $L_3(1,2,2)$ is also realized. Now consider the following amalgamation diagram.

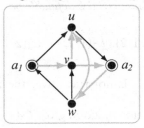

This involves triangle types
$$L_3(1,1,2),\ L_3(1,2,2),\ L_3(2,1,2),\ L_3(2,2,1),\ C_3(2,2,1)$$

and so has a completion in Γ, which must have $a_1 \xrightarrow{2} a_2$. Thus $L_3(2,2,2)$ is realized.

Now consider the following amalgamation diagram.

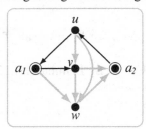

This involves triangle types $L_3(1,1,2)$, $L_2(1,2,2)$, $L_3(2,2,1)$, and $L_3(2,2,2)$ and so has a completion in Γ, which must have $a_2 \xrightarrow{2} a_1$. Thus $C_3(2,2,2)$ is realized.

This completes the proof. □

21.4.6. Proof of Proposition 21.4.2.

LEMMA 21.4.23. *Suppose that Γ is a homogeneous primitive 3-constrained 2-multi-tournament. Suppose that the triangle type $C_3(1,1,1)$ is forbidden and the triangle types*

$$L_3(1,1,2), \qquad\qquad L_3(2,2,1),$$
$$C_3(1,1,2), \qquad\qquad C_3(2,2,1), \qquad\qquad C_3(2,2,2)$$

are realized. Then the triangle types $L_3(1,2,1)$, $L_3(2,1,1)$ are also realized.

PROOF. By symmetry it suffices to prove that $L_3(1, 2, 1)$ is realized. Suppose on the contrary that this type is forbidden. Then easily $L_3(1, 1, 1)$ is realized. Consider the following amalgamation diagram.

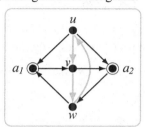

This involves the triangle types $L_3(1, 1, 2)$, $L_3(2, 1, 1)$, $C_3(1, 1, 2)$, and $C_3(2, 2, 2)$ and has no completion in Γ. Therefore the triangle type $L_3(2, 1, 1)$ must also be forbidden.

In particular our hypotheses are now preserved by reversal of the orientations of both 2-types.

Consider the following amalgamation diagram.

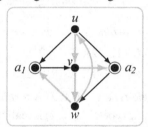

This involves the triangle types $L_3(1, 1, 2)$, $L_3(2, 2, 1)$, $L_3(2, 1, 2)$, $C_3(1, 1, 2)$, $C_3(2, 2, 1)$, and $C_3(2, 2, 2)$, and has no completion in Γ. Therefore the triangle type $L_3(2, 1, 2)$ must also be forbidden. By symmetry, $L_3(1, 2, 2)$ is also forbidden.

Then the following amalgamation diagram yields a contradiction.

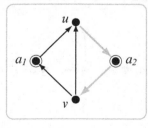

This proves the lemma.   □

LEMMA 21.4.24. *Suppose that Γ is a homogeneous 3-constrained 2-multi-tournament for which the triangle type $C_3(1, 1, 1)$ is forbidden and the triangle* *types*

$$L_3(1, 1, 2), L_3(2, 1, 1), L_3(2, 2, 1)$$
$$C_3(1, 1, 2), C_3(2, 2, 1)$$

are realized. Then the triangle type $L_3(2, 2, 2)$ is realized.

PROOF. The following diagram involves only the realized triangle types.

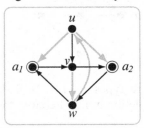

Any allowable completion will contain a triangle of type $L_3(2, 2, 2)$. The lemma follows.   □

LEMMA 21.4.25. *Suppose that Γ is a homogeneous primitive 3-constrained 2-multi-tournament not associated with a free amalgamation class. Suppose that the triangle type $C_3(1, 1, 1)$ is forbidden and the triangle types*

$$L_3(1, 1, 2), \ L_3(2, 2, 1)$$
$$C_3(2, 2, 1), \ C_3(2, 2, 2)$$

are realized. Then the triangle type $C_3(1, 1, 2)$ is forbidden.

PROOF. Suppose the triangle type $C_3(1, 1, 2)$ is realized. Then Lemma 21.4.23 shows that triangle types $L_3(1, 2, 1)$ and $L_3(2, 1, 1)$ are realized. Then Lemma 22.5.2 shows that the triangle type $L_3(2, 2, 2)$ is also realized.

As Γ is not associated with a free amalgamation class, at least one of the triangle types $L_3(1, 2, 2)$ or $L_3(2, 1, 2)$ must be forbidden. By symmetry we may suppose that $L_3(1, 2, 2)$ is forbidden.

Consider the following amalgamation diagram.

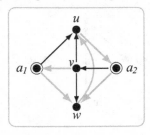

This involves the triangle types $L_3(1, 1, 2)$, $L_3(1, 2, 1)$ $L_3(2, 1, 1)$, $L_3(2, 2, 1)$, $C_3(1, 1, 2)$, $C_3(2, 2, 2)$ and has no completion in Γ. This is a contradiction.   □

LEMMA 21.4.26. *Suppose that* Γ *is a 3-constrained homogeneous primitive 2-multi-tournament with no non-trivial $\emptyset$-definable partial order, with triangle types*

$$C_3(1, 1, 1), \ C_3(1, 1, 2)$$

forbidden, and triangle type $C_3(2, 2, 2)$ realized.

Then after a suitable identification of the languages, Γ is isomorphic to entry #8, #10, or #12 of Table 21.1.

PROOF. If $L_3(2, 2, 1)$ is forbidden then Proposition 21.4.10 applies and $\Gamma \cong \mathbb{S}(4)$ under a suitable identification of the languages.

If $C_3(2, 2, 1)$ is forbidden then after reversing the 2-type $\xrightarrow{1}$ Proposition 21.4.10 applies, and again $\Gamma \cong \mathbb{S}(4)$ under a suitable identification of the languages.

So we suppose that triangle types

$$L_3(2, 2, 1), \ C_3(2, 2, 1)$$

are realized.

At this point the finite case may be removed by inspection, so we suppose Γ is infinite.

If triangle type $L_3(1, 1, 1)$ is forbidden then Proposition 21.4.19 applies and gives an identification with entry #10.

If triangle type $L_3(1, 1, 1)$ is realized then Proposition 21.4.21 applies and gives an identification with entry #8. □

PROOF OF PROPOSITION 21.4.2. We assume that the triangle type $C_3(1, 1, 1)$ is realized and the triangle type $C_3(2, 2, 2)$ is forbidden. We aim at one of entries 8, 10, 12 in our table. We consider four cases.

(a) If triangle type $L_3(2, 2, 1)$ is forbidden then Proposition 21.4.10 applies, and Γ is isomorphic to $\mathbb{S}(4)$.

(a') If $C_3(2, 2, 1)$ is forbidden then after reversing $\xrightarrow{2}$ we fall under the same case.

(b) If triangle type $C_3(1, 1, 2)$ is forbidden then Lemma 21.4.26 applies.

(b') If $L_3(1, 1, 2)$ is forbidden then after reversing a 2-type the same lemma applies.

Thus we arrive at the following case.

Triangle type $C_3(1, 1, 1)$ is forbidden.

Triangle types $L_3(1, 1, 2)$, $L_3(2, 2, 1)$, $C_3(1, 1, 2)$, $C_3(2, 2, 1)$, and $C_3(2, 2, 2)$ are realized.

But by Lemma 21.4.25, this case does not arise. □

21.4.7. Entry 9: Proposition 21.4.3. We now aim at the following.

PROPOSITION 21.4.3 (NPO$_{3c}$: #9). *Suppose that* Γ *is a homogeneous primitive 3-constrained 2-multi-tournament with no nontrivial* $\emptyset$-*definable partial order, and which does not have free amalgamation. Suppose that the triangle types* $C_3(1, 1, 1)$ *and* $C_3(2, 2, 2)$ *are both realized in* Γ.

Then up to a change of language, Γ *is determined by the constraints in entry #9 of our table: that is, triangle types* $C_3(1, 1, 2)$ *and* $L_3(1, 1, 1)$ *are forbidden and all other triangle types are realized.*

LEMMA 21.4.27. *Suppose that* Γ *is a homogeneous primitive 2-multi-tournament with no* $\emptyset$-*definable partial order, and not associated with a free amalgamation class. Suppose that the triangle types*

$$C_3(1, 1, 1), \ C_3(2, 2, 2)$$

are realized and the triangle type $L_3(1, 2, 1)$ *is forbidden.*
Then the triangle types

$$L_3(1, 1, 1), L_3(1, 1, 2), C_3(1, 1, 2)$$

are realized.

PROOF. We may deal with the finite case by inspection so we suppose Γ is infinite. In that case as $L_3(1, 2, 1)$ is forbidden, the triangle type $L_3(1, 1, 1)$ is realized.

By symmetry with respect to reversal of the type $\xrightarrow{2}$, it suffices now to show that $L_3(1, 1, 2)$ is realized. So suppose toward a contradiction that it is forbidden.

CLAIM 1. *The triangle type* $C_3(1, 1, 2)$ *is realized.*

If this type is forbidden then by primitivity the type $L_3(2, 1, 1)$ is realized. But then the following diagram forces $C_3(1, 1, 2)$.

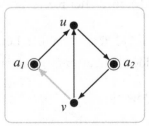

This proves the claim.

CLAIM 2. *The triangle type* $C_3(2, 2, 1)$ *is realized.*

The following diagram forces triangle type $L_3(2, 1, 1)$.

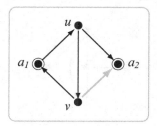

Then the following forces $C_3(2, 1, 1)$.

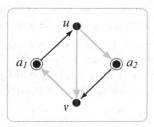

This proves the claim.
Now the following diagram forces $L_3(1, 1, 2)$.

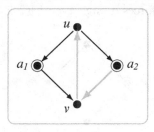

$\square$

LEMMA 21.4.28. *Suppose that* Γ *is a 3-constrained homogeneous primitive 2-multi-tournament with no* $\emptyset$*-definable partial order, and not associated with a free amalgamation class. Suppose that the triangle types*

$$C_3(1, 1, 1), \ C_3(2, 2, 2)$$

are realized and the triangle type $L_3(1, 2, 1)$ *is forbidden.*
Then the triangle type

$$L_3(2, 1, 1)$$

is forbidden.

PROOF. Consider the following amalgamation diagram.

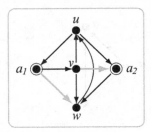

With the exception of triangle type $L_3(2, 1, 1)$ the triangle types involved in the factors are realized by hypothesis or by Lemma 21.4.27.

As this diagram has no completion in Γ, the lemma follows.   □

LEMMA 21.4.29. *Suppose that Γ is a homogeneous primitive 2-multi-tournament with no $\emptyset$-definable partial order, and not associated with a free amalgamation class. Suppose that the triangle types $C_3(1, 1, 1)$ and $C_3(2, 2, 2)$ are realized and $L_3(1, 2, 1)$, $L_3(2, 1, 1)$ are forbidden. Then the triangle types*

$$L_3(2, 2, 1), \ C_3(2, 2, 1)$$

are realized.

PROOF. By symmetry it suffices to deal with $L_3(2, 2, 1)$.
The following diagram forces triangle type $L_3(2, 2, 1)$ to be realized.

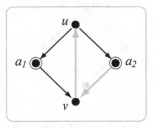

          □

LEMMA 21.4.30. *Suppose that Γ is a 3-constrained homogeneous primitive 2-multi-tournament with no $\emptyset$-definable partial order, and not associated with a free amalgamation class. Suppose that the triangle types*

$$C_3(1, 1, 1), \ C_3(2, 2, 2)$$

are realized and triangle type $L_3(1, 2, 1)$ is forbidden.
Then the triangle types

$$L_3(1, 2, 2), \ L_3(2, 1, 2)$$

are forbidden.

PROOF. By symmetry it suffices to deal with $L_3(1, 2, 2)$.
Consider the following amalgamation diagram.

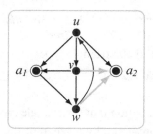

The triangle types involved are $L_3(1,1,1)$, $L_3(1,1,2)$, $L_3(1,2,2)$, and $C_3(1,1,1)$. Other than $L_3(1,2,2)$, these are afforded by hypothesis and by Lemma 21.4.27. As this diagram has no completion in Γ the lemma follows. □

LEMMA 21.4.31. *Suppose that Γ is a 3-constrained homogeneous primitive 2-multi-tournament with no $\emptyset$-definable partial order, and not associated with a free amalgamation class. Suppose that the triangle types $C_3(1,1,1)$ and $C_3(2,2,2)$ are realized. Then the triangle types*

$$L_3(1,2,1), \qquad L_3(1,2,2), \qquad L_3(2,1,1), \qquad L_3(2,1,2)$$

are realized.

PROOF. By symmetry, it suffices to treat the case of $L_3(1,2,1)$.

Suppose that the triangle type $L_3(1,2,1)$ is forbidden.

Then the triangle types $L_3(2,1,1)$, $L_3(1,2,2)$, $L_3(2,1,2)$ are also forbidden by Lemma 21.4.28 and 21.4.30. It follows easily that triangle type $L_3(2,2,2)$ is realized.

The following amalgamation diagram involves triangle types $L_3(1,1,1)$, $L_3(1,1,2)$, $L_3(2,2,1)$, $L_3(2,2,2)$, $C_3(1,1,2)$, and $C_3(2,2,1)$. All of these except $L_3(2,2,2)$ are afforded by Lemmas 21.4.28 and 21.4.29.

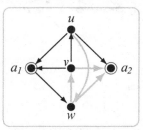

Any completion of this diagram contains a triangle of type $L_3(1,2,1)$ or $L_3(2,1,1)$, giving a contradiction. The lemma follows. □

LEMMA 21.4.32. *Suppose that Γ is a 3-constrained homogeneous primitive 2-multi-tournament with no $\emptyset$-definable partial order. Suppose that the triangle types*

$$C_3(1,1,1), \ C_3(2,2,2)$$

are realized and the triangle type $C_3(1, 1, 2)$ is forbidden.
   Then the triangle type $L_3(1, 1, 2)$ is realized.

PROOF. By Lemma 21.4.31 the triangle types

$$L_3(1, 2, 1) \qquad L_3(1, 2, 2) \qquad L_3(2, 1, 1) \qquad L_3(2, 1, 2)$$

are realized.

Suppose toward a contradiction that the triangle type $L_3(1, 1, 2)$ is forbidden. Then our hypotheses are preserved by reversal of the type $\xrightarrow{2}$.

Consider the following amalgamation diagram.

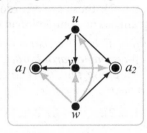

This has no completion, and involves triangle types $L_3(1, 2, 1)$, $L_3(2, 1, 1)$, $L_3(2, 1, 2)$, $L_3(2, 2, 1)$, and $C_3(1, 1, 1)$. Therefore the triangle type $L_3(2, 2, 1)$ must be forbidden.

Reversing the type $\xrightarrow{2}$, the triangle type $C_3(2, 2, 1)$ is also forbidden.

Consider the following amalgamation diagram.

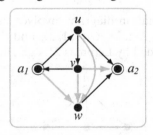

This involves triangle types

$$L_3(1, 2, 1), \ L_3(1, 2, 2), \ L_3(2, 1, 1), \ L_3(2, 1, 2), \ C_3(1, 1, 1),$$

but has no completion in Γ. This is a contradiction.   □

LEMMA 21.4.33. *Suppose that Γ is a 3-constrained homogeneous primitive 2-multi-tournament with no $\emptyset$-definable partial order. Suppose that the triangle types*

$$C_3(1, 1, 1), \ C_3(2, 2, 2)$$

are realized, and the triangle type $C_3(1, 1, 2)$ is forbidden. Then the triangle type $L_3(2, 2, 1)$ is realized.

PROOF. Suppose toward a contradiction that the triangle type

$$L_3(2, 2, 1)$$

is forbidden.

Then switching the labels 1, 2 and reversing the 2-type $\xrightarrow{2}$ in Lemma 21.4.32 we find that the triangle type

$$C_3(2, 2, 1)$$

is realized.

CLAIM 1. *The triangle type $L_3(1, 1, 1)$ is realized.*

If we suppose the contrary then the triangle type $L_3(2, 2, 2)$ is realized. Consider the following amalgamation diagram.

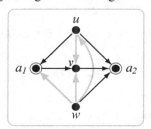

This involves the triangle types $L_3(1, 1, 2)$, $L_3(2, 1, 1)$, $L_3(2, 1, 2)$, and $L_3(2, 2, 2)$. A completion must realize triangle type $L_3(1, 1, 1)$. This proves the claim.

CLAIM 2. *Triangle type $L_3(2, 2, 2)$ is realized.*

Consider the following amalgamation diagram.

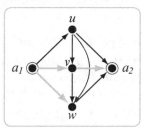

This involves the triangle types $L_3(1, 1, 1)$, $L_3(1, 1, 2)$, $L_3(1, 2, 1)$, and $L_2(2, 1, 2)$. The unique completion has $a_1 \xrightarrow{2} a_2$ and thus $L_3(2, 2, 2)$ is realized.

Now we consider an amalgamation diagram with 6 points.

This does not involve the triangle types $C_3(1, 1, 2)$ or $L_3(2, 2, 1)$ and all others are realized at this point. As there is no completion we arrive at a contradiction. Thus the type $L_3(2, 2, 1)$ is realized.

This completes the proof. □

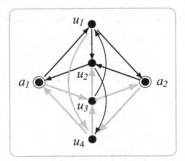

LEMMA 21.4.34. *Suppose that* Γ *is a homogeneous primitive 2-multi-tournament with no $\emptyset$-definable partial order. Suppose that the triangle types*

$$L_3(1,1,2),\ C_3(2,2,2)$$

are realized, and the triangle type $C_3(1,1,2)$ *is forbidden.*
  Then the triangle type

$$C_3(2,2,1)$$

is realized.

  PROOF. The following diagram forces $C_3(2,2,1)$ to be realized.

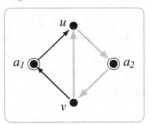

The factors are triangles of types $L_3(1,1,2)$ and $C_3(2,2,2)$.   □

  LEMMA 21.4.35. *Suppose that* Γ *is a 3-constrained homogeneous primitive 2-multi-tournament with no $\emptyset$-definable partial order, and not associated with a free amalgamation class. Suppose that the triangle types*

$$C_3(1,1,1),\ C_3(2,2,2)$$

are realized, and the triangle type $C_3(1,1,2)$ *is forbidden.*
  Then the triangle type $L_3(1,1,1)$ *is forbidden, and all remaining triangle types are realized.*

  PROOF. Lemmas 21.4.31–21.4.34 dispose of everything other than the monochromatic triangles of types $L_3(1,1,1)$, $L_3(2,2,2)$. So we need to show that the former is forbidden and the latter is realized.

  CLAIM 1. *The triangle type* $L_3(2,2,2)$ *is realized.*

The following amalgamation diagram involves the triangle types $L_3(1, 2, 1)$, $L_3(1, 2, 2)$, $L_3(2, 1, 1)$, $L_3(2, 1, 2)$, and $C_3(2, 2, 1)$.

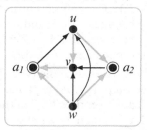

Any completion has a 2-arc between a_1 and a_2 and thus realizes the triangle type $L_3(2, 2, 2)$. This proves the claim.

CLAIM 2. *The triangle type $L_3(1, 1, 1)$ is forbidden.*

The following amalgamation diagram does not involve the triangle type $C_3(1, 1, 2)$ and has no completion in Γ.

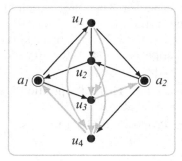

As the only triangle type involved which could be forbidden is $L_3(1, 1, 1)$, this proves the claim.

This completes the proof of the lemma.   □

PROOF OF PROPOSITION 21.4.3. We suppose that Γ realizes $C_3(1, 1, 1)$ and $C_3(2, 2, 2)$.

By Lemma 21.4.31, Γ realizes the triangle types $L_3(1, 2, 1)$, $L_3(1, 2, 2)$, $L_3(2, 1, 1)$, $L_3(2, 1, 2)$. As Γ can forbid at most one of $L_3(1, 1, 1)$ or $L_3(2, 2, 2)$ and is not a free amalgamation class, it must forbid one of the remaining triangle types

$$L_3(1, 1, 2), \qquad L_3(2, 2, 1), \qquad C_3(1, 1, 2), \qquad C_3(2, 2, 1).$$

Up to symmetry, that is, allowing the labels $1, 2$ to be swapped and the orientation of $\xrightarrow{2}$ to be reversed, we may suppose therefore that Γ does not realize the type $C_3(1, 1, 2)$. Then Lemma 21.4.35 completes the proof.   □

21.5. Proof of completeness (3-constrained case)

PROOF OF PROPOSITION 21.3.1. Γ is infinite, primitive, homogeneous, and we assume it has no $\emptyset$-definable linear order and is not associated with a free amalgamation class. We suppose also that Γ is 3-constrained. We claim then that it is listed in Table 21.1 as one of entries 6–12.

- If there is a transitive 2-type we assume the type $\xrightarrow{1}$ is transitive. We apply Proposition 21.3.2, proved on p. 125. As explained following the statement of that proposition, this leads to entries #6,7 in the table.
- If there is no transitive 2-type, then as we assume there is no $\emptyset$-definable linear order, we arrive at the case in which there is no $\emptyset$-definable partial order. Depending on which of the triangle types $C_3(1,1,1)$, $C_3(2,2,2)$ is realized, one of Propositions 21.4.1, 21.4.2, or 21.4.3 applies.
  - If $C_3(1,1,1)$, $C_3(2,2,2)$ are both forbidden then Proposition 21.4.1, proved on p. 134, leads to entry #11 ($\widetilde{\mathbb{S}(3)}$).
  - If exactly one of $C_3(1,1,1)$ and $C_3(2,2,2)$ is forbidden, we may assume that $C_3(1,1,1)$ is forbidden. Then Proposition 21.4.2, proved on p. 147, leads to one of the entries #8, 10, or 12 ($\mathbb{S}(4)$).
  - If both $C_3(1,1,1)$ and $C_3(2,2,2)$ are realized then Proposition 21.4.3, proved on p. 155, applies, and leads to entry #9.

Thus all cases are covered.   □

PROOF OF PROPOSITION 21.1.2. This is Proposition 21.2.1, proved on p. 118, plus Proposition 21.3.1, proved above.   □

As noted at the outset, Proposition 21.1.1 follows.

Chapter 22

HOMOGENEOUS 2-MULTI-TOURNAMENTS:
FORBIDDEN TRIANGLES; CONCLUSION

22.1. Introduction

Having identified the 3-constrained homogeneous 2-multi-tournaments, the next question is the extent to which the constraints on triangles found in a general homogeneous 2-multi-tournament agree with these. In other words, if one looks only at the constraints on triangles, do they already define an amalgamation class? (If so, the next question is whether the original structure is determined by a set of Henson constraints relative to the larger 3-constrained class.)

In the case in which $\overset{1}{\to} \cup \overset{2}{\to}$ is required to be a linear order (up to a change of language) this question is settled by the explicit classification in Part I. Here we discuss the possible patterns of triangle constraints in the remaining cases. Recall that this kind of analysis is the customary prelude to an identification of exceptional examples (in the case of homogeneous directed graphs, the exotic example $\mathcal{P}(3)$ was found at this stage).

We aim here at the following partial result.

PROPOSITION 22.1.1. *Let Γ be an infinite, primitive, homogeneous 2-multi-tournament not associated with a free amalgamation class.*

If Γ has a $\emptyset$-definable linear order then it is found in the classification in Part I, and is listed in Table 18.1, group II (entries 2–5).

If Γ has no $\emptyset$-definable linear order, then either

1. *the set of forbidden triangles in Γ defines one of the known 3-constrained 2-multi-tournaments listed in Table 21.1, or*
2. *one of the following four cases applies.*
    (a) *Triangle types $C_3(1, 1, 1)$ and $C_3(2, 2, 2)$ are forbidden and all other triangle types are realized.*
    (b) *Triangle types $C_3(1, 1, 1)$ and $L_3(2, 2, 1)$ are forbidden and all other triangle types are realized.*
    (c) *Triangle type $C_3(1, 1, 2)$ is forbidden and all other triangle types are realized.*

(d) *Triangle types* $C_3(1, 1, 2)$ *and* $L_3(2, 2, 1)$ *are forbidden and all other triangle types are realized.*

We conjecture that the four exceptional cases envisioned here do not arise.

We proceed as follows. Proposition 22.2.1 deals with the case in which there is a definable partial order, with no exceptional cases left over. The remaining cases are subdivided as follows (assuming throughout that there is no definable partial order).

1. If triangle types $C_3(1, 1, 1)$ and $C_3(2, 2, 2)$ are forbidden, we arrive either at the restrictions associated with $\widetilde{\mathbb{S}(3)}$, or the first of our exceptional cases (Proposition 22.3.1).
2. If triangle types $C_3(1, 1, 1)$ and $L_3(2, 2, 1)$ are forbidden, while triangle type $C_3(2, 2, 2)$ is realized, we arrive either at the restrictions associated with $\mathbb{S}(4)$, or the second of our exceptional cases (Proposition 22.4.1).
3. If the triangle types $C_3(1, 1, 1)$ and $C_3(1, 1, 2)$ are forbidden, while the triangle types $L_3(1, 1, 1)$, $L_3(2, 2, 1)$, $C_3(2, 2, 1)$ are realized, then all other triangle types are realized, as in entry #8 (Proposition 22.6.1).
4. If the triangle types $C_3(1, 1, 1)$ and $C_3(2, 2, 2)$ are realized and the triangle type $C_3(1, 1, 2)$ is forbidden, then all other triangle types are realized except possibly $L_3(2, 2, 1)$ and $L_3(1, 1, 1)$ (Proposition 22.7.1).

In §22.8 we check that these cases cover everything necessary to derive Proposition 22.1.1—but actually, we also need a generalization of Lemma 21.4.25, originally proved only in the 3-constrained case. This is given as Lemma 22.5.3. The proof becomes considerably more elaborate at this level of generality.

We make extensive use of explicit amalgamation arguments throughout.[5]

We continue to make use of Notation 20.1.1 in conjunction with the abbreviated notations 1, 2, 1^{op}, 2^{op} for $\xrightarrow{1}$, $\xrightarrow{2}$, $\xleftarrow{1}$ and $\xleftarrow{2}$ respectively.

There is also some further use of Notation 21.1.3 below. Notably, in the proof of Proposition 22.2.1 following, we have both v^2 (as in Notation 20.1.1) and L_3^2 (as in Notation 21.1.3).

22.2. The transitive case

PROPOSITION 22.2.1. *Suppose that* Γ *is a homogeneous primitive 2-multi-tournament in which the relation* $\xrightarrow{1}$ *is transitive (that is, the triangle types* $C_3(1, 1, 1)$, $C_3(1, 1, 2)$, *and* $L_3(1, 1, 2)$ *are forbidden) and there is no* $\emptyset$-*definable linear order.*

---

[5] One can imagine substantial generalizations of the these results to richer languages, possibly with less ad hoc proofs, but even if that materializes it is very probable that such arguments would begin with very similar lemmas proved by amalgamation arguments (or, indeed, by some sort of induction with the type of argument carried out here providing part of the base of the induction).

Then all other triangle types are realized, except possibly $L_3(2,2,1)$ or $C_3(2,2,1)$; at least one of these is realized as well. Thus, up to a change of language, the triangle constraints are as in entry #6 or #7 of Table 21.1.

PROOF. With the exception of the triangle types

$$L_3(2,2,1), \quad C_3(2,2,1),$$

this is all covered by Proposition 21.3.2. That is, we begin with the following information.

Triangle types $L_3(1,1,2)$, $C_3(1,1,1)$, $C_3(1,1,2)$ are forbidden.

Triangle types $L_3(1,2,1)$, $L_3(1,2,2)$, $L_3(2,1,1)$, $L_3(2,2,1)$,

$$L_3(2,2,2), \text{ and } C_3(2,2,2) \text{ are realized.}$$

We assume that the conclusion fails.

Triangle types $L_3(2,2,1)$, $C_3(2,2,1)$ are forbidden.

In view of Lemma 21.3.10 we also have the following.

The configurations $(IC_3)^2$ and $(C_3I)^2$ are forbidden.

Consider the following amalgamation diagram.

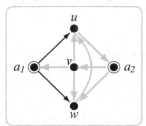

Any completion will contain a triangle of type $L_3(2,2,1)$ or $C_3(2,2,1)$, so it suffices now to embed the two factors into Γ.

The factor omitting a_1 is the unique completion of the following diagram, since $(C_3I)^2$ is forbidden.

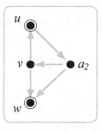

So it suffices to embed the factor omitting a_2, or the configuration obtained by reversing the orientation of all its arcs, into Γ.

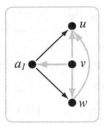

This may be viewed as a triangle of type $L_3(1, 2, 1)$ sitting in v^2. The reversal may be viewed as a triangle of type $L_3(2, 1, 1)$ sitting in $v^{2^{\text{op}}}$.

By Lemma 21.3.7 we may take a configuration aLb in Γ with

$$a \xrightarrow{1} L \xrightarrow{1} b; \qquad\qquad L \cong L_3^2.$$

We may then take $u \in \Gamma$ with

$$u \xrightarrow{2} a; \qquad\qquad b \xrightarrow{2} u.$$

We find that u is related to L by 2-arcs. Thus we may find $c_1, c_2 \in L$ with $u \xrightarrow{2} c_1, c_2$ or $c_1, c_2 \xrightarrow{2} u$, and this gives either $u \xrightarrow{2} L_3(1, 2, 1)$ or $L_3(2, 1, 1) \xrightarrow{2} u$, as required.

This completes the proof. $\qquad\qquad\qquad\qquad\qquad\qquad\qquad\qquad\qquad\square$

22.3. No definable p.o.: around $\widetilde{\mathbb{S}(3)}$

We take up the more complex case in which Γ has no non-trivial $\emptyset$-definable partial order. The main division is according to how many of the oriented 3-cycles C_3^1 and C_3^2 are realized, with the case in which exactly one is realized branching off into three separate cases; in total, these correspond to the five entries 8–12 of our table.

We aim now at the following.

PROPOSITION 22.3.1 (NPO: $\widetilde{\mathbb{S}(3)}$). *Suppose that Γ is an infinite homogeneous primitive 2-multi-tournament in which there is no $\emptyset$-definable partial order and triangles of types $C_3(1, 1, 1)$, $C_3(2, 2, 2)$, are forbidden. Then one of the following two possibilities occurs.*

(a) *Up to a change of language, a triangle of type $L_3(1, 2, 1)$ is forbidden and all other triangle types are realized, as in the 3-constrained case #11 (the structure $\widetilde{\mathbb{S}(3)}$).*

(b) *All other triangle types are realized in Γ.*

The second alternative should not occur, but that is a difficult point as the relevant amalgamation diagrams in the 3-constrained case have factors of order 5.

Still, we get this far by making a close study of forbidden structures of order 4.

22.3.1. Notation: 2-multi-tournaments of order 4. We will need to consider which configurations of order 4 embed into Γ. We adopt the following notation for such configurations.

NOTATION 22.3.2. L_4, C_4, IC_3, and C_3I denote the four tournaments of order 4, namely the transitive tournament, the tournament containing a 4-cycle, and the two tournaments consisting of a vertex which either dominates or is dominated by a 3-cycle, respectively. We label the vertices of these tournaments according to the following conventions.

L_4: $1 \to 2 \to 3 \to 4$.
C_4: $1 \to 2 \to 3 \to 4 \to 1$ and $1 \to 3, 2 \to 4$.
IC_3: $1 \to (2,3,4)$, with $2 \to 3 \to 4 \to 2$ taken in any cyclic order.
C_3I: $(2,3,4) \to 1$, similarly.

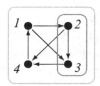

FIGURE 6. The tournament C_4:
$$1 \to (2,3) \to 4 \to 1.$$

We view any 2-multi-tournament as a tournament with arcs colored. Accordingly we use the notation $T(i_1 i_2 i_3 i_4 i_5 i_6)$ for a 2-multi-tournament of order 4, where T is the underlying tournament, and $i_1, \ldots, i_6$ are the colors (i.e., the labels $1, 2$) according to some fixed arrangement of pairs of vertices. We use the following convention.

L_4: $(1,2), (2,3), (3,4); (1,3), (2,4); (1,4)$.
C_4: $(1,2), (2,3), (3,4), (4,1); (1,3), (2,4)$.
IC_3: $(1,2), (1,3), (1,4); (2,3), (3,4), (4,2)$.
C_3I: $(2,1), (3,1); (4,1); (2,3), (3,4), (4,2)$.

FIGURE 7. The tournament $C_4(1,2,3,4;5,6)$

Furthermore, we apply this notation at present only to 2-multi-tournaments which omit the 3-cycles $C_3(1,1,1)$, $C_3(1,1,2)$, $C_3(2,2,2)$. Therefore, in cases where the underlying tournament is IC_3 or C_3I, the 3-cycle involved must be $C_3(2, 2, 1)$; and we adopt the additional convention that

$$2 \xrightarrow{2} 3 \xrightarrow{2} 4 \xrightarrow{1} 2$$

in this case.

We therefore have the following notations, with some added punctuation for the sake of legibility.

- $L_4(i_1 i_2 i_3; j_1 j_2; k)$;
- $C_4(i_1 i_2 i_3 i_4; j_1 j_2)$;
- $IC_3(ijk; 221)$; $C_3I(ijk; 221)$.

One must also keep track of the symmetries available (reversing the orientation of both 2-types). These transform the configurations as follows.

$L_4(i_1 i_2 i_3; j_1 j_2; k) \leftrightarrow L_4(i_3 i_1 i_2; j_2 j_1; k)$.
$C_4(i_1 i_2 i_3 i_4; j_1 j_2) \leftrightarrow C_4(i_3 i_2 i_1 i_4; j_2 j_1)$.
$IC_3(ijk; 221) \leftrightarrow C_3I(kji; 221)$.

22.3.2. The triangle types $L_3(1, 1, 2)$, $L_3(2, 2, 1)$, $C_3(1, 1, 2)$, and $C_3(2, 2, 1)$. We aim now to generalize Lemma 21.4.6, dropping the hypothesis of 3-constraint. That is, we wish to show the following.

LEMMA 22.3.3. *Suppose that Γ is a homogeneous primitive 2-multi-tournament in which there is no nontrivial $\emptyset$-definable partial order and triangles of types $C_3(1, 1, 1)$, $C_3(2, 2, 2)$ are forbidden. Then the triangle types*

$$L_3(1, 1, 2), \quad L_3(2, 2, 1), \quad C_3(1, 1, 2), \quad C_3(2, 2, 1)$$

are realized in Γ.

These four types correspond under changes of language, that is, switching labels 1, 2 or reversing the orientation of 2-types. So it will suffice to prove that a particular one of these types is realized in Γ, and we focus on $C_3(1, 1, 2)$.

Thus we will spend some time exploring the consequences of forbidding triangle types $C_3(1, 1, 1)$, $C_3(2, 2, 2)$, and $C_3(1, 1, 2)$. To begin with, we review the relevant results from §21.4.1.

LEMMA 22.3.4. *Suppose that Γ is a homogeneous primitive 2-multi-tournament in which there is no nontrivial $\emptyset$-definable partial order and triangles of types $C_3(1, 1, 1)$, $C_3(1, 1, 2)$, and $C_3(2, 2, 2)$ are forbidden. Then the triangle types*

$$L_3(1, 1, 1), \quad L_3(1, 1, 2), \quad L_3(1, 2, 1), \quad L_3(1, 2, 2),$$
$$L_3(2, 1, 1), \quad L_3(2, 1, 2), \quad C_3(2, 2, 1)$$

are realized in Γ.

PROOF. This is stated a little more generally in Lemma 21.4.4. $\square$

LEMMA 22.3.5. *Suppose that* Γ *is a homogeneous primitive 2-multi-tournament in which there is no nontrivial $\emptyset$-definable partial order and triangles of types*

$$C_3(1,1,1), \ C_3(1,1,2), \ C_3(2,2,2)$$

are forbidden.

Then the following hold.

1. *The configurations*

$$C_4(1112;22), \ C_4(2121;22)$$

   embed into Γ.

2. *The configurations*

$$C_4(1212;22), \ L_4(111;22;2), \ L_4(121;21;2), \ L_4(121;12;2),$$
$$C_4(2221;22)$$

   do not embed into Γ.

PROOF.

CLAIM 1. *The configuration* $C_4(1112;22)$ *embeds into* Γ.

This configuration may be viewed as an amalgamation determining the type of the pair of vertices $(1,3)$. The factors $C_3(221)$ and $L_3(112)$ embed in Γ and the unique possible amalgam is $C_4(1112;22)$.

CLAIM 2. *The configuration* $C_4(1212;22)$ *does not embed in* Γ.

Assuming the contrary, make the following amalgamation.

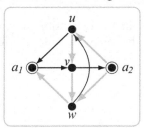

The factors are $C_4(1112;22)$ and $C_4(1212;22)$ (amalgamated over vertices 2,4,1 or 4,1, 3 respectively). The parameters u, v, w forbid (a_1, a_2) to have type $\xrightarrow{1}, \xleftarrow{1}$ or $\xleftarrow{2}$, and $\xrightarrow{2}$, respectively.

This gives a contradiction and proves the claim.

CLAIM 3. *The configuration* $C_4(2121;22)$ *embeds in* Γ.

We view this configuration as an amalgam with the type of the pair of vertices $(1,3)$ to be determined. The desired type is $1 \xrightarrow{2} 3$.

It is also possible to complete the amalgamation by $3 \xrightarrow{2} 1$, but this would give $C_4(1212;22)$.

CLAIM 4. *The configurations*

$$L_4(111; 22; 2), \quad L_4(121; 21; 2), \quad L_4(121; 12; 2)$$

do not embed in Γ.

If $L_4(111; 22; 2)$ or $L_4(121; 21; 2)$ embeds in Γ we make use of one of the following amalgamations.

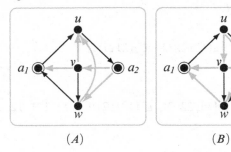

(A)   (B)

These have factors

$$L_4(111; 22; 2), \quad C_4(2121; 22), \quad C_4(2121; 22), \quad L_4(121; 21; 2)$$

respectively. In each case the parameters u, v, w prevent any completion by a 2-type for a_1, a_2.

This disposes of the first two configurations, and by symmetry $L_4(121; 12; 2)$ is also forbidden.

CLAIM 5. *One of the configurations* $L_4(221; 12; 2)$ *or* $L_4(211; 22; 2)$ *embeds into* Γ.

We view $L_4(221; 12; 2)$ as an amalgamation diagram with the type of $(1, 2)$ to be determined. This has the following four solutions.

$a_1 \xrightarrow{1} a_2$: $L_4(121; 12; 2)$;

$a_1 \xleftarrow{1} a_2$: $L_4(111; 22; 2)$;

$a_1 \xrightarrow{2} a_2$: $L_4(221; 12; 2)$;

$a_1 \xleftarrow{2} a_2$: $L_4(211; 22; 2)$.

As the first two are forbidden, the claim follows.

CLAIM 6. *The configuration* $C_4(2221; 22)$ *does not embed in* Γ.

Assuming the contrary, and invoking the previous claim, we may use one of the following amalgamations (A, B).

These have factors

$$L_4(221; 12; 2), \quad C_4(2221; 22), \quad L_4(211; 22; 2), \quad C_4(2221; 22)$$

respectively, and neither has a consistent completion. This gives a contradiction and proves the claim.

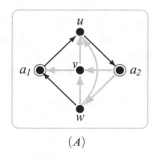
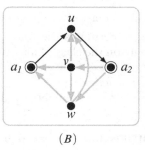

(A)   (B)

This completes the proof of the lemma.   □

LEMMA 22.3.6. *Suppose that Γ is a homogeneous primitive 2-multi-tournament in which there is no nontrivial $\emptyset$-definable partial order and triangles of types $C_3(1,1,1)$, $C_3(1,1,2)$, $C_3(2,2,2)$, are forbidden. Then $L_3(222)$ embeds into Γ.*

PROOF. We treat $L_4(111;22;2)$ as an amalgamation problem with the type of the pair of vertices $(3,4)$ to be determined.

The possible solutions are

$3 \xrightarrow{1} 4$: $L_4(111;22;2)$

$3 \xleftarrow{1} 4$: $L_4(121;21;2)$

$3 \xrightarrow{2} 4$ or $3 \xleftarrow{2} 4$: $(1,3,4)$ has type $L_3(222)$.

By the previous lemma the first two possibilities are excluded.   □

At this point we have shown that when $C_3(1,1,1)$, $C_3(1,1,2)$ and $C_3(2,2,2)$ are forbidden, then all triangle types other than the three forbidden 3-cycles, and possibly the configuration $L_3(221)$, embed into Γ.

We summarize the information obtained up to this point in the following table.

	Forbidding $C_3(1,1,1)$, $C_3(1,1,2)$, $C_3(2,2,2)$	
	3-types	*4-types*
Forbidden	$C_3(111)$, $C_3(112)$, $C_3(222)$	$C_4(1212;22)$, $C_4(2221;22)$, $L_4(111;22;2)$, $L_4(121;21;2)$, $L_4(121;12;2)$
Realized	$C_3(221)$, all L_3 except possibly $L_3(221)$	$C_4(1112;22)$, $C_4(2121;22)$

Now we deal with the remaining triangle type, $L_3(221)$. If this triangle type is omitted, then we have an additional symmetry to take into account.

Remark 22.3.7. Let τ operate on 2-types by replacing $\xrightarrow{1}$ by $\xleftarrow{2}$, and $\xrightarrow{2}$ by $\xrightarrow{1}$. Then τ permutes the four non-trivial 2-types cyclically, and acts correspondingly on the class of all 2-multi-tournaments.

The constraint set $C_3(1, 1, 1)$, $C_3(1, 1, 2)$, $C_3(2, 2, 2)$, $L_3(2, 2, 1)$ is invariant setwise under the action of τ.

Note that τ^2 is the symmetry previously considered: reversal of both orientations.

The next argument is not easy to find.

LEMMA 22.3.8. *Suppose that* Γ *is a homogeneous primitive 2-multi-tournament in which there is no nontrivial $\emptyset$-definable partial order and triangles of types*

$$C_3(111), \quad C_3(112), \quad C_3(222)$$

are forbidden.

Then the triangle type $L_3(221)$ *embeds in* Γ.

PROOF. For the duration of the argument we assume toward a contradiction that $L_3(221)$ is forbidden.

CLAIM 1. *The configuration* $L_4(112; 22; 2)$ *embeds into* Γ.

We view this configuration as an amalgamation diagram with the type of vertices $(1, 4)$ to be determined.

Our additional constraint $L_3(221)$ prevents $1 \xrightarrow{1} 4$ and thus the only possible completion has $a_1 \xrightarrow{2} a_2$, giving $L_4(112; 22; 2)$.

CLAIM 2. *The configuration* $C_4(2121; 22)$ *does not embed into* Γ.

Assuming the contrary, use the following amalgamation.

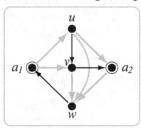

The factors are $C_4(2121; 22)$ and $L_4(122; 22; 2)$, and there is no consistent completion.

CLAIM 3. *The configuration* $C_4(1212; 22)$ *embeds into* Γ.

View this as an amalgamation diagram with the type of vertices $(1, 3)$ to be determined.

Any completion has a 2-arc between vertex 1 and 3.

– If $1 \xrightarrow{2} 3$: this is the desired configuration $C_4(1212; 22)$.

– If $3 \xrightarrow{2} 1$: this would be $C_4(2121; 22)$ which is forbidden by the previous claim.

Now consider the following amalgamation.

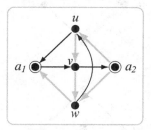

This has factors $C_4(1112; 22)$ and $C_4(1212; 22)$, and has no completion. This gives a contradiction, and proves the lemma. □

LEMMA 22.3.9. *Suppose that Γ is a homogeneous primitive 2-multi-tournament in which there is no nontrivial $\emptyset$-definable partial order and triangles of types $C_3(111)$, $C_3(112)$, $C_3(222)$ are forbidden. Then all other triangle types embed in Γ.*

PROOF. Lemmas 22.3.4, 22.3.6, and 22.3.8 cover all cases. □

After these lengthy preparations we can eliminate this subcase.

PROOF OF LEMMA 22.3.3. We suppose the triangle types $C_3(1, 1, 1)$, $C_3(2, 2, 2)$ are forbidden and we aim to show that the triangle types

$$L_3(1, 1, 2), \; L_3(2, 2, 1), \; C_3(1, 1, 2), \; C_3(2, 2, 1)$$

are realized.

As these four triangle types correspond under reversing one 2-type or switching the labels $1, 2$, in the contrary case we may suppose toward a contradiction that

the triangle type $C_3(112)$ is forbidden.

Then by Lemma 22.3.9 all triangle types other than

$$C_3(1, 1, 1), \; C_3(1, 1, 2), \; C_3(2, 2, 2)$$

embed into Γ.

CLAIM 1. *The configuration $C_4(1112; 22)$ embeds into Γ.*

Make the following amalgamation, with factors $L_3(221)$, $L_3(112)$.

The only completion has $a_1 \xrightarrow{2} a_2$ and is of type $C_4(1112; 22)$. This proves the claim.

CLAIM 2. *The configuration $C_4(1212; 22)$ is forbidden.*

The following amalgamation diagram has no completion consistent with our restrictions.

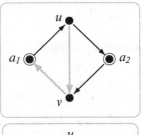

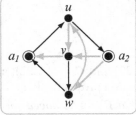

The two factors of this diagram have the forms $C_4(1112; 22)$ and $C_4(1212; 22)$, and the former embeds in Γ, so the latter must be forbidden.

CLAIM 3. *The configuration $C_4(2121; 22)$ embeds into Γ.*

Make the following amalgamation, with factors $C_3(221)$, $L_3(122)$.

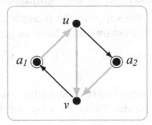

The possible completions have $a_1 \xrightarrow{2} a_2$ or $a_1 \xleftarrow{2} a_2$, giving one of the configurations $C_4(2121; 22)$ or $C_4(1212; 12)$. As the latter is forbidden, the former is realized.

CLAIM 4. *The configurations $L_4(111; 22; 2)$ and $L_4(121; 12; 2)$ are forbidden.*

We consider the following pair of amalgamation diagrams. Neither has a completion consistent with our restrictions.

The factors are as follows.

A: $L_4(111; 22; 2)$ and $C_4(2121; 22)$;
B: $L_4(121; 12; 2)$ and $C_4(2121; 22)$.

As $C_4(2121; 22)$ embeds into Γ, the other two factors, $L_4(111; 22; 2)$ and $L_4(121; 12; 2)$, must be forbidden.

CLAIM 5. *At least one of the configurations $L_4(211; 22; 2)$ or $L_4(221; 12; 2)$ embeds into Γ.*

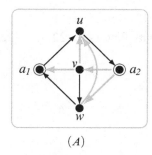 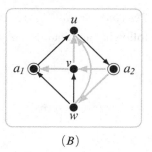

                 (A) (B)

Make the following amalgamation, with factors $L_3(212)$, $L_3(112)$.

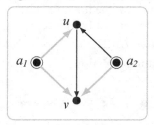

This has the following four possible completions.

$$L_4(111; 22; 2), \; L_4(121; 12; 2), \; L_4(211; 22; 2), \; L_4(221; 12; 2).$$

By the previous claim, the first two possibilities are excluded and the last two remain.

CLAIM 6. *The configuration $C_4(2221; 22)$ is forbidden.*

We consider the following pair of amalgamation diagrams. Neither has a completion consistent with our restrictions.

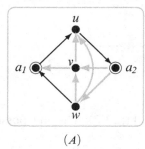 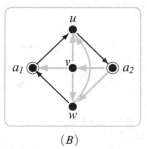

                 (A) (B)

The factors are as follows.

A: $L_4(221; 12; 2)$ and $C_4(2221; 22)$;
B: $L_4(211; 22; 2)$ and $C_4(2221; 22)$.

By the previous claim, at least one of the factors of type L_4 here embeds into Γ, so the factor $C_4(2221; 22)$ must be forbidden.

CLAIM 7. *The configuration $C_4(1222; 12)$ embeds into Γ.*

Make the following amalgamation, with factors $C_3(221)$, $L_3(222)$.

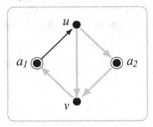

The possible completions have the form $C_4(1222; 12)$ and $C_4(2221; 22)$. As the latter is forbidden, the former embeds into Γ.

CLAIM 8. *The configuration $L_4(221; 12; 2)$ is forbidden.*

Consider the following amalgamation.

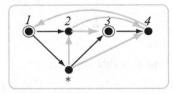

The factors are $C_4(1222; 12)$ and $L_4(221; 12; 2)$. The only possible completion is $C_4(1212; 22)$.

By Claim 2, this diagram has no completion in Γ. But the factor $C_4(1222; 12)$ embeds into Γ, so the factor $L_4(221; 12; 2)$ is forbidden.

CLAIM 9. *The configuration $L_4(211; 22; 2)$ is forbidden.*

Consider the following amalgamation.

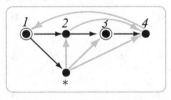

The factors are $C_4(1222; 12)$ and $L_4(211; 22; 2)$. The only possible completion is $L_4(121; 22; 2)$.

By Claim 4 the diagram has no completion in Γ. But $C_34(1222; 12)$ embeds in Γ, so the configuration $L_4(211; 22; 2)$ is forbidden.

Now to conclude the proof of the lemma, notice that Claims 5, 8, and 9 give a contradiction.   □

22.3.3. Around $\widetilde{\mathbb{S}(3)}$, continued. We are now dealing with the following constraints on triangles.

Forbidden	Realized
$C_3(111)$, $C_3(222)$	$L_3(112)$, $L_3(221)$,
	$C_3(112)$, $C_3(221)$

According to Lemma 21.4.7, when we consider the triangle types $L_3(1, 2, 1)$, $L_3(1, 2, 2)$, $L_3(2, 1, 2)$, $L_3(2, 1, 1)$, up to a change of language we arrive at the following four cases.

  (a) Triangle types $L_3(1, 2, 2)$ and $L_3(2, 1, 2)$ are forbidden, while $L_3(1, 2, 1)$ and $L_3(2, 1, 1)$ are realized.
  (a′) Triangle types $L_3(1, 2, 1)$ and $L_3(2, 1, 2)$ are forbidden, while $L_3(1, 2, 2)$ and $L_3(2, 1, 1)$ are realized.
  (b) Triangle type $L_3(1, 2, 1)$ is forbidden, while $L_3(1, 2, 2)$, $L_3(2, 1, 1)$, $L_3(2, 1, 2)$ are realized.
  (c) Triangle types $L_3(1, 2, 1)$, $L_3(1, 2, 2)$, $L_3(2, 1, 1)$, and $L_3(2, 1, 2)$ are all realized.

We eliminate case (a').

LEMMA 22.3.10. *Suppose that Γ is a homogeneous primitive 2-multi-tournament in which there is no nontrivial $\emptyset$-definable partial order, and that Γ satisfies the following constraints.*

Triangle types $L_3(1, 2, 1)$, $L_3(2, 1, 2)$, and $C_3(1, 1, 1)$ are forbidden.

Triangle types $L_3(1, 1, 2)$ and $L_3(2, 1, 1)$ are realized.

Then Γ is finite.

PROOF. Assume toward a contradiction that Γ is infinite.
Take $a \in \Gamma$ and set

$$\Gamma^+ = a^1; \qquad\qquad \Gamma^- = a^{1^{\mathrm{op}}}.$$

Then (Γ^-, Γ^+) is homogeneous as a partitioned 2-multi-tournament. Since Γ is primitive, infinite, and $\aleph_0$-categorical, the two parts are infinite.

CLAIM 1. $(\Gamma^+, \xrightarrow{1}) \cong (\mathbb{Q}, <)$. *In particular the triangle type $L_3(1, 1, 1)$ is realized.*

As the triangle types $L_3(1, 2, 1)$ and $C_3(1, 1, 1)$ are not realized, Γ^+ is an infinite homogeneous $\xrightarrow{1}$-tournament without oriented 3-cycles. The claim follows.

CLAIM 2. Γ^- *realizes both $\xrightarrow{1}$ and $\xrightarrow{2}$.*

In Γ this simply means that triangle types $L_3(1, 1, 1)$ and $L_3(2, 1, 1)$ are realized.

CLAIM 3. *For $u \in \Gamma^-$ there is a unique $u^+ \in \Gamma^+$ satisfying*

$$u \xrightarrow{2} u^+.$$

Existence: This means that the triangle type $L_3(1, 1, 2)$ is realized in Γ.
Uniqueness: This holds since the triangle type $L_3(2, 1, 2)$ is not realized in Γ.
The claim follows.

Now write

$$f : \Gamma^- \to \Gamma^+$$

for the function $f(u) = u^+$. Define $x \sim y$ on Γ^- by

$$f(x) = f(y).$$

As Γ^+ realizes one pair of anti-symmetric 2-types and Γ^- realizes two such pairs, the relation $\sim$ is $\overset{i}{\sim}$ for some $i = 1$ or 2 (notation as in Definition 20.3.1). We now adopt the notation $\{i, j\} = \{1, 2\}$.

Consider $u, v \in \Gamma^-$ with $f(u) \overset{1}{\to} f(v)$. Γ^- is a composition in which there are arcs of type j either from $u/\sim$ to $v/\sim$ or the reverse. This gives a triangle of type $L_3(j, i, j)$ in Γ, contradicting our assumptions. $\qquad \square$

22.3.4. Reduction to cases (a, c). Our goal in the present section is the following. The labels on the cases are carried over from Lemma 21.4.7.

LEMMA 22.3.11. *Suppose that Γ is an infinite homogeneous primitive 2-multi-tournament in which there is no nontrivial $\emptyset$-definable partial order and triangles of types*

$$C_3(1, 1, 1), \ C_3(2, 2, 2)$$

are forbidden.

Then up to a permutation of the language one of the following cases applies.

(*a*) *Triangle types*

$$L_3(1, 2, 2), \ L_3(2, 1, 2)$$

are forbidden, while triangle types

$$L_3(1, 2, 1), \ L_3(2, 1, 1)$$

are realized.

(*c*) *Triangle types*

$$L_3(1, 2, 1), \ L_3(1, 2, 2), \ L_3(2, 1, 1), \ L_3(2, 1, 2)$$

are all realized.

In view of Lemmas 22.3.3, 21.4.7, and 22.3.10 there is only one case other than the two listed that comes into consideration at this point, namely the case in which the triangle type $L_3(1, 2, 1)$ is forbidden, while triangle types

$$L_3(1, 1, 2), \ L_3(2, 2, 1), \ C_3(1, 1, 2), \ C_3(2, 2, 1),$$
$$L_3(1, 2, 2), \ L_3(2, 1, 1), \ L_3(2, 1, 2)$$

are realized. So we restate our objective more concretely.

LEMMA 22.3.12. *Suppose that Γ is a homogeneous primitive 2-multi-tourna-ment in which there is no nontrivial $\emptyset$-definable partial order and triangles of types*

$$L_3(1,2,1), \ C_3(1,1,1), \ C_3(2,2,2)$$

are forbidden and all other triangle types, except possibly the types $L_3(1,1,1)$ or $L_3(2,2,2)$, are realized. Then Γ is finite.

We use our customary notation for 2-multi-tournaments of order four (Notation 22.3.2). First we require the following.

LEMMA 22.3.13. *Suppose that Γ is a homogeneous primitive 2-multi-tourna-ment in which there is no nontrivial $\emptyset$-definable partial order. Suppose that triangle types*

$$L_3(1,2,1), \ C_3(1,1,1), \ C_3(2,2,2)$$

are forbidden and all other triangle types, except possibly types $L_3(1,1,1)$ or $L_3(2,2,2)$, are realized.

Then the configuration $C_4(1112;11)$ is forbidden.

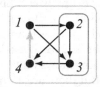

PROOF. We suppose toward a contradiction that the specified configuration is realized in Γ.

Fix $a \in \Gamma$ and define

$$\Gamma^+ = a^1; \qquad\qquad \Gamma^- = a^{2^{\mathrm{op}}}.$$

Then (Γ^+, Γ^-) is a homogeneous partitioned 2-multi-tournament.

CLAIM 1. *Γ^+ is a $\xrightarrow{1}$-tournament isomorphic with $(\mathbb{Q}, <)$.*

As Γ does not realize the triangle type $L_3(1, 2, 1)$, Γ^+ is a $\xrightarrow{1}$-tournament. As $C_4(1112; 11)$ is realized, $L_3(111)$ is realized and thus Γ^+ is non-trivial. As $C_3(1, 1, 1)$ is forbidden, Γ^+ is transitive. Thus

$$\Gamma^+ \cong (\mathbb{Q}, <).$$

Now for $u \in \Gamma^-$ define

$$I_u = \{v \in \Gamma^+ \mid v \xrightarrow{1} u\}.$$

Schematically, the situation is as follows.

$$\Gamma^- \ \xrightarrow{2} \ a \ \xrightarrow{1} \ \Gamma^+$$
$$u \qquad\qquad\qquad \xleftarrow{1} \ I_u$$

Claim 2. I_u *is an interval in* Γ^+.

If $v_1, v_2 \in I_u$ and $v_1 \xrightarrow{1} v \xrightarrow{1} v_2$, then as $u, v \in v_1^1$ we have either $u \xrightarrow{1} v$ or $v \xrightarrow{1} u$. As $v \xrightarrow{1} v_2 \xrightarrow{1} u$ we do not have $u \xrightarrow{1} v$, so $v \xrightarrow{1} u$. Thus the interval (v_1, v_2) in Γ^+ lies in I_u and the claim follows.

Claim 3. I_u *is a non-trivial bounded interval in* Γ^+.

As we assume the configuration $C_4(1112; 11)$ is realized in Γ, it follows that I_u contains at least two points.

As the triangle types $L_3(2, 1, 1)$ and $L_3(2, 1, 2)$ are realized in Γ, we can find elements v_1, v_2 of Γ^+ with $u \xrightarrow{1} v_1$ and $u \xrightarrow{2} v_2$. We claim that v_1, v_2 are respectively an upper and a lower bound for I_u.

So take $w \in I_u$. Since $w \xrightarrow{1} u \xrightarrow{1} v_1$ we do not have $v_1 \xrightarrow{1} w$, so $w \xrightarrow{1} v_1$. Similarly if $w \xrightarrow{1} v_2$ then (w, u, v_2) has type $L_3(1, 2, 1)$, a contradiction. Thus the claim holds.

Claim 4. *If* $v_1 \xrightarrow{1} v_1' \xrightarrow{1} v_2' \xrightarrow{1} v_2$ *in* Γ^+ *then there is* $u \in \Gamma^-$ *with* $(v_1', v_2') \subseteq I_u \subseteq (v_1, v_2)$.

As the configuration $(a v_1 v_1' v_2' v_2)$ realizes a unique 5-type it suffices to find a single case in which this holds. For this, start with I_u and then choose v_1, v_1', v_2, v_2' correspondingly. This proves the claim.

Now we may reach a contradiction. By considering the relation of I_{u_1} to I_{u_2} for $u_1, u_2 \in \Gamma^-$, the previous claim shows that there are at least six non-trivial 2-types realized in Γ^- over a. But there are at most four such 2-types.

This contradiction proves the lemma.   □

Proof of Lemma 22.3.12. We suppose toward a contradiction that Γ is infinite. Then for $a \in \Gamma$, a^1 is infinite. As $L_3(1, 2, 1)$ is not realized, the triangle type $L_3(1, 1, 1)$ is realized.

Claim 1. *The configuration* $C_4(2112; 11)$ *is forbidden.*

Consider the following amalgamation diagram.

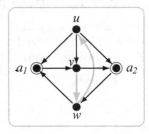

As $C_3(2, 2, 2)$ and $L_3(1, 2, 1)$ are forbidden this has no completion and therefore one of the factors is not realized in Γ.

But the factor $a_2 u v w$ is the unique amalgam of $u v w$ with $u a_2 w$ over u, w, since $L_3(1, 2, 1)$ is forbidden, and these triangles have types $C_3(2, 2, 1)$ and

$C_3(1,1,2)$, both of which are realized in Γ. So it is the first factor a_1uvw, of type $C_4(2112;11)$, which must be forbidden. This proves the claim.

CLAIM 2. *The configuration $IC_3(211;112)$ is realized.*

Consider the following amalgamation diagram, with factors $C_3(1,1,2)$ and $L_3(1,1,1)$.

By the previous claim, the completion cannot have $1 \xrightarrow{1} 2$. As $L_3(1,2,1)$ is forbidden, the completion must have $1 \xrightarrow{2} 2$ or $2 \xrightarrow{2} 1$. If $1 \xrightarrow{2} 2$ then the configuration is $C_4(1112;11)$, contradicting Lemma 22.3.13.

So $2 \xrightarrow{2} 1$ and the configuration is $IC_3(211;112)$. The claim is proved.

Now we use a variant of our first amalgamation diagram, as shown below.

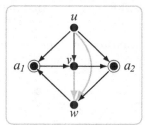

This time our previous claim gives the factor a_1uvw, and the second factor is the unique amalgam of uvw with ua_2w over uw, involving triangle types $L_3(1,2,2)$ and $L_3(1,1,2)$. So this gives a contradiction. $\qquad\square$

22.3.5. Case (a). Now the four cases introduced in Lemma 21.4.7 under the assumption that triangle types C_3^1 and C_3^2 are forbidden have been reduced to two, namely cases (a, c) as originally enumerated.

(a)

Forbidden		Realized	
$L_3(1,2,2)$	$L_3(2,1,2)$	$L_3(1,2,1)$	$L_3(2,1,1)$

(c)

Forbidden	Realized		
$L_3(1,2,1)$	$L_3(1,2,2)$	$L_3(2,1,1)$	$L_3(2,1,2)$

We deal next with case (a). Our goal is the following.

PROPOSITION 22.3.14. *Suppose that Γ is an infinite homogeneous primitive 2-multi-tournament in which there is no nontrivial $\emptyset$-definable partial order and triangles of types*

$$L_3(1,2,2),\ L_3(2,1,2),\ C_3(1,1,1),\ C_3(2,2,2)$$

are forbidden. Then the triangle type $L_3(1, 1, 1)$ *is also forbidden, and all other triangle types are realized, as in* $\mathbb{S}(3)$, *entry #11, Table 21.2.*

Lemma 22.3.15. *Suppose that* Γ *is a homogeneous primitive 2-multi-tournament in which there is no nontrivial $\emptyset$-definable partial order and triangles of types*

$$L_3(1, 2, 2), \ L_3(2, 1, 2), \ C_3(1, 1, 1), \ C_3(2, 2, 2)$$

are forbidden. Then the triangle type $L_3(1, 1, 1)$ *is also forbidden.*

Proof. We recall that triangles of types

$$L_3(1, 1, 2), \ L_3(2, 2, 1), \ C_3(1, 1, 2), \ C_3(2, 2, 1)$$

must be realized (Lemma 22.3.3). We suppose toward a contradiction that the triangle type $L_3(1, 1, 1)$ is also realized.

We consider the partitioned 2-multi-tournament $\mathbb{T}_a = (a^{2^{\mathrm{op}}}, a^2)$ relative to some basepoint a. By our assumptions the components of $\mathbb{T}_a$ are transitive tournaments ordered by $\xrightarrow{2}$ with type $\mathbb{Q}$, so that the type $\xrightarrow{1}$ occurs only as a cross type.

Claim 1. *Every configuration on four points with the following structure embeds into* Γ.

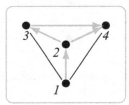

We make the following amalgamation.

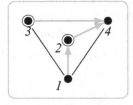

For any orientation of the 1-arcs, the factors are realized in Γ.

As $L_3(122)$ is forbidden, the unique completion has $2 \xrightarrow{2} 3$. This proves the claim.

Now we work toward the following diagram, in which the orientation of the arc between u and w will need to be determined.

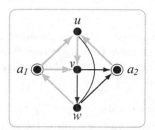

If we can construct the factors of this diagram in Γ then we arrive at a contradiction, since any completion will involve one of the triangle types $L_3(1, 2, 2)$ or $L_3(2, 1, 2)$.

To determine the orientation of the 1-arc between u and w, we perform the following amalgamation with factors $C_3(221)$ and $L_3(111)$.

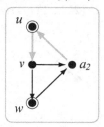

This then determines the matching factor required for our final amalgamation.

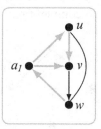

This has the structure referred to in our first claim, so embeds in Γ.

Thus we construct the desired diagram and we reach a contradiction.   □

PROOF OF PROPOSITION 22.3.14. By Lemmas 22.3.11 and 22.3.3 the triangle types

$$L_3(1, 1, 2), \ L_3(1, 2, 1), \ L_3(2, 1, 1), \ L_3(2, 2, 1), \ C_3(1, 1, 2), \ C_3(2, 2, 1)$$

are realized.

By Lemma 22.3.15 the triangle type $L_3(1, 1, 1)$ is forbidden, and as Γ is infinite the triangle type $L_3(2, 2, 2)$ is realized.

This completes the proof.   □

22.3.6. Case (c).

PROPOSITION 22.3.16. *Suppose that* Γ *is an infinite homogeneous primitive 2-multi-tournament in which there is no nontrivial $\emptyset$-definable partial order and triangles of types $C_3(111)$, $C_3(222)$, are forbidden, and the triangle types*

$$L_3(1,2,1), \ L_3(1,2,2), \ L_3(2,1,1), \ L_3(2,1,2)$$

are all realized. Then all other triangle types are realized as well.

PROOF. Lemma 22.3.3 covers all remaining types other than $L_3(1,1,1)$ and $L_3(2,2,2)$. As Γ is infinite at least one of these triangle types is realized in Γ. By symmetry, it suffices to treat the case in which

$$L_3(2,2,2) \text{ embeds in } \Gamma; \text{ and}$$

$$L_3(1,1,1) \text{ does not.}$$

CLAIM 1. *The configuration $C_3(2222;11)$ is realized.*

It suffices to view this configuration as an amalgamation diagram with the type of $(1,3)$ to be determined (if $3 \xrightarrow{1} 1$ then the vertices should be relabeled).

CLAIM 2. *The configuration $C_4(2121;22)$ is forbidden.*

The following diagram has no completion.

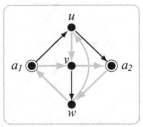

The factors are of the form $C_4(2121;22)$ and $C_4(2222;11)$, and as the latter embeds in Γ, the former must be forbidden.

CLAIM 3. *The configuration $L_4(212;21;2)$ embeds in Γ.*

Fix the vertex *1* as basepoint and view this configuration as a triangle of type $L_3(121)$ embedding in I^2. Since triangles of type $L_3(111)$ are assumed forbidden, this configuration is realized.

CLAIM 4. *The configuration $C_3 I(221; 221)$ embeds in Γ.*

Label the vertices $1, 2, 3, 4$ with $1, 2, 3$ the copy of $C_3(221)$. View the configuration as an amalgamation diagram with the type of $(1, 4)$ to be determined. Since triangle type $L_3(1, 1, 1)$ is forbidden, the vertex 3 forces $1 \xrightarrow{2} 4$ or $1 \xleftarrow{2} 4$. The vertex 2 determines the orientation.

CLAIM 5. *The configuration $C_4(1212; 22)$ embeds into Γ.*

We make use of the following amalgamation.

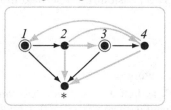

The factor omitting vertex 3 has type $C_3 I(221; 221)$ while the factor omitting vertex 1 has type $L_4(212; 21; 2)$. Thus both embed in Γ.

In the completed diagram, the pair $(1, 3)$ realize type 2 in some orientation. The orientation $3 \xrightarrow{2} 1$ gives the configuration $C_4(2121; 22)$, contradiction Claim 2. So we are left with $1 \xrightarrow{2} 3$ and $C_4(1212; 22)$.

CLAIM 6. *The configuration $L_4(222; 11; 2)$ is realized.*

Our hypotheses allow the symmetry in which $\xrightarrow{1}$ is replaced by its reversal. Then Claim 5 translates into the stated claim.

Now we may conclude. We consider the following diagram, which has no completion.

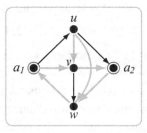

Its factors are of types

$$C_4(1212; 22), \quad L_4(222; 11; 2),$$

which we have shown embed in Γ. Thus we have reached a contradiction. □

22.3.7. Proof of Proposition 22.3.1.

PROOF OF PROPOSITION 22.3.1, P. 160. Under the hypotheses of the proposition, Lemma 22.3.11 describes two possible conclusions, which following our earlier analysis are called cases (a) and (c). Proposition 22.3.14 states that in case (a) we arrive at the same triangle constraints as in $\overline{\mathbb{S}(3)}$. Proposition 22.3.16 states that in case (c) all triangle types are realized other than $C_3(1, 1, 1)$ and $C_3(2, 2, 2)$.

Thus we arrive at the two possibilities described in Proposition 22.3.1.   □

22.4. Around $\mathbb{S}(4)$

Now we aim at the following.

PROPOSITION 22.4.1 ($\mathbb{S}(4)$ dichotomy). *Suppose that Γ is a homogeneous primitive 2-multi-tournament with no nontrivial $\emptyset$-definable partial order, in which the triangle types*

$$L_3(2, 2, 1), \ C_3(1, 1, 1)$$

are forbidden and the triangle type $C_3(2, 2, 2)$ is realized.

Then either Γ is isomorphic to $\mathbb{S}(4)$, or Γ realizes all triangle types other than $L_3(2, 2, 1)$ and $C_3(1, 1, 1)$

22.4.1. Positive constraints.

LEMMA 22.4.2. *Suppose that Γ is a homogeneous primitive 2-multi-tournament with no nontrivial $\emptyset$-definable partial order. Suppose that the triangle types $C_3(1, 1, 1)$ and $L_3(2, 2, 1)$ are forbidden and the triangle type $C_3(2, 2, 2)$ is realized.*

Then all triangle types realized in $\mathbb{S}(4)$ are realized in Γ, namely

$$L_3(1, 1, 1), \ L_3(1, 1, 2), \ L_3(1, 2, 2), \ L_3(2, 1, 2), \ C_3(2, 2, 1), \ C_3(2, 2, 2).$$

PROOF. Lemmas 21.4.11 and 21.4.12 apply and give triangle types

$$L_3(1, 1, 2), \ L_3(1, 2, 2), \ L_3(2, 1, 2).$$

Since we have type $C_3(2, 2, 2)$ by hypothesis we are concerned only with the two triangle types $L_3(1, 1, 1)$ and $C_3(2, 2, 1)$.

CLAIM 1. *The triangle type $L_3(1, 1, 1)$ is realized.*

We consider the following amalgamation diagram.

The only possible completion has $a_1 \xrightarrow{1} a_2$ and thus a_1, u, a_2 has type $L_3(1, 1, 1)$.

The factor $a_1 uvw$ is the only possible amalgam of $a_1 uv$ and $a_1 vw$ over $a_1 v$ which does not itself contain $L_3(1, 1, 1)$, and the factor $a_2 uvw$ is the only possible amalgam of $a_2 uv$ and uvw over uv which does not itself contain

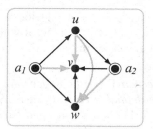

$L_3(1, 1, 1)$. So by considering these amalgams we either get $L_3(1, 1, 1)$ directly or the factors of our original diagram.

The triangles occurring as subfactors in this construction have types $L_3(1, 2, 2)$, $L_3(1, 1, 2)$, and $L_3(2, 1, 1)$. The first two of these are known to be realized in Γ, and if the type $L_3(2, 1, 1)$ is not realized in Γ, then easily the type $L_3(1, 1, 1)$ is realized.

This completes the proof of the claim in all cases.

CLAIM 2. *The triangle type $C_3(2, 2, 1)$ is realized.*

We consider the following amalgamation diagram, with the orientation of the 2-arc between v and w to be determined.

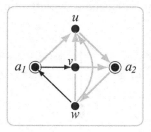

All allowable completions involve a triangle of type $C_3(2, 2, 1)$ so it suffices to find a form of this diagram whose factors embed in Γ.

We view the factor a_2uvw as an amalgamation diagram with the type of (v, w) to be determined. Then this is necessarily a 2-arc. We consider the corresponding factor a_1uvw with (v, w) realizing the same 2-type. This will result from one of the following two amalgamation diagrams.

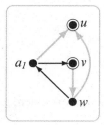 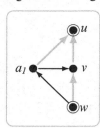

More precisely, if the triangle type $C_3(2, 2, 1)$ does not occur in the completion of the relevant diagram, then the required factor will be obtained.

Thus the claim holds in any case, and with this the lemma is proved.   □

22.4.2. A first dichotomy.

LEMMA 22.4.3. *Suppose that Γ is a homogeneous primitive 2-multi-tournament with no nontrivial $\emptyset$-definable partial order. Suppose that the triangle types $C_3(1, 1, 1)$ and $L_3(2, 2, 1)$ are forbidden and the triangle type $C_3(2, 2, 2)$ is realized. Then one of the following holds.*

1. *Γ is isomorphic with $\mathbb{S}(4)$.*
2. *Γ realizes the triangle types $L_3(1, 2, 1)$ and $L_3(2, 1, 1)$.*

PROOF. By Lemma 22.4.2, Γ realizes every triangle type realized in $\mathbb{S}(4)$. If in addition the four triangle types

$$L_3(1, 2, 1),\ L_3(2, 1, 1),\ L_3(2, 2, 2),\ C_3(1, 1, 2) \qquad (*)$$

are not realized in Γ then Γ and $\mathbb{S}(4)$ realize the same triangle types. In this case, by Lemma 21.2.11, Γ must be isomorphic to $\mathbb{S}(4)$.

Let us assume now that triangle types $L_3(1, 2, 1)$ and $L_3(2, 1, 1)$ are not both realized. By symmetry, we may suppose that the triangle type $L_3(1, 2, 1)$ is forbidden.

Then the following diagram has no completion in Γ.

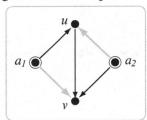

As this is an amalgam of triangles of types $L_3(1, 1, 2)$ and $L_3(2, 1, 1)$ and the triangle type $L_3(1, 1, 2)$ is realized in Γ, the triangle type $L_3(2, 1, 1)$ is not realized.

Similarly, consideration of the diagram

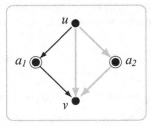

shows that triangle type $L_3(2, 2, 2)$ must be forbidden, and consideration of the diagram

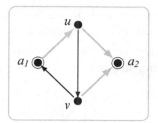

shows that the triangle type $C_3(1, 1, 2)$ must be forbidden.

Thus the four triangle types listed in $(*)$ are forbidden, and Γ is isomorphic to $\mathbb{S}(4)$. □

While we will not eliminate the second case here, we will show that it corresponds to a definite pattern of forbidden triangles, namely: $L_3(2, 2, 1)$ and $C_3(1, 1, 1)$ forbidden and all others realized.

22.4.3. The bad case. For reference we restate our point of departure as a lemma, which we use without explicit mention.

LEMMA 22.4.4. *Suppose that Γ is a homogeneous primitive 2-multi-tournament with no nontrivial $\emptyset$-definable partial order for which the triangle types*

$$L_3(2, 2, 1), \ C_3(1, 1, 1)$$

are forbidden and the triangle type

$$C_3(2, 2, 2)$$

is realized, and that Γ is not isomorphic to $\mathbb{S}(4)$. Then Γ realizes all other triangle types except possibly $C_3(1, 1, 2)$ and $L_3(2, 2, 2)$. That is, Γ realizes the following triangle types.

$$L_3(1, 1, 1), \ L_3(1, 1, 2), \ L_3(1, 2, 1), \ L_3(1, 2, 2), \ L_3(2, 1, 1), \ L_3(2, 1, 2),$$
$$C_3(2, 2, 1), \ C_3(2, 2, 2).$$

PROOF. Lemmas 22.4.2 and 22.4.3. □

LEMMA 22.4.5. *Suppose that Γ is a homogeneous primitive 2-multi-tournament with no nontrivial $\emptyset$-definable partial order for which the triangle types*

$$L_3(2, 2, 1), \ C_3(1, 1, 1)$$

are forbidden and the triangle type

$$C_3(2, 2, 2)$$

is realized, and that Γ is not isomorphic to $\mathbb{S}(4)$. Then Γ realizes the triangle type $C_3(1, 1, 2)$.

PROOF. We make a lengthy analysis of the contrary case. Thus we suppose that the triangle types

$$L_3(2, 2, 1), \ C_3(1, 1, 1), \ C_3(1, 1, 2)$$

are forbidden and that all others, with the possible exception of $L_3(2, 2, 2)$, are realized. We first address that possible exception.

CLAIM 1. *The triangle type $L_3(2, 2, 2)$ is realized.*

We consider the following diagram.

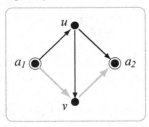

The factors have type $L_3(1, 1, 2)$ and $L_3(1, 2, 1)$ and any completion in Γ will have $a_1 \xrightarrow{2} a_2$. The claim follows.

CLAIM 2. *The configurations*

$$IC_3(2, 2, 2; 2, 2, 1), \quad C_4(2, 2, 2, 1; 2, 2) \tag{22.1}$$

are forbidden.

We make use of the following amalgamation diagram, with the 2-arc between u and w having either orientation.

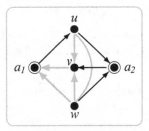

Regardless of the orientation of the 2-arc between u and w, the factor a_2uvw must embed in Γ: with $u \xrightarrow{2} w$ it is the unique completion of the amalgam of a_1uw with a_2vw over a_2w, and with $w \xrightarrow{2} u$ it is the unique completion of the amalgam of a_2uv with a_2uw over a_2u. Thus both forms of the factor a_1uvw are forbidden and this gives the claim.

CLAIM 3. *The configuration*

$$IC_3(1, 2, 2; 2, 2, 1) \tag{22.2}$$

is realized.

Consider the following amalgamation diagram.

Since the triangle type $C_3(1, 1, 2)$ is not realized, in the completion we cannot have $1 \xrightarrow{1} 2$. By Claim 2, the relations $2 \xrightarrow{2} 1$ and $1 \xrightarrow{2} 2$ are also forbidden,

Thus in the completion $2 \xrightarrow{1} 1$ must hold, and hence $IC_3(1, 2, 2; 2, 2, 1)$ is realized, as claimed.

CLAIM 4. *The configuration $L_4(1, 1, 1; 1, 2; 2)$ is forbidden.*

Consider the following amalgamation diagram, with factors $a_1 uvw$ and $a_2 uvw$ of the forms $IC_3(1, 2, 2; 2, 2, 1)$ and $L_4(1, 1, 1; 1, 2; 2)$, respectively.

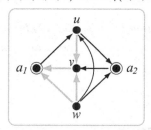

The diagram has no completion consistent with the forbidden triangles, and the first factor of the diagram is afforded by Claim 3, so the claim follows.

CLAIM 5. *The configuration $L_4(1, 1, 1; 1, 1; 2)$ is forbidden.*

Consider the following amalgamation diagram, which has no allowable completion consistent with the forbidden triangles.

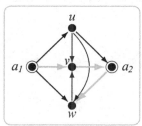

The factor $a_2 uvw$ is the unique amalgam of the triangles $a_2 uw$ and $a_2 vw$ over $a_2 w$, and hence embeds into Γ.

As the factor $a_1 uvw$ has the form $L_4(1, 1, 1; 1, 1; 2)$, the claim follows.

After all these preparations, to reach a contradiction it suffices to consider the following diagram.

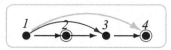

The completion has type $L_4(1, 1, 1; 1, 1; 2)$ or $L_4(1, 1, 1; 1, 2, 2)$, which contradicts Claim 4 or 5.   □

Now we can complete the proof of our dichotomy.

PROOF OF PROPOSITION 22.4.1. By Lemmas 22.4.4 and 22.4.5 we may suppose that Γ forbids the triangle types

$$L_3(2, 2, 1), \ C_3(1, 1, 1)$$

and realizes all others, with the possible exception of $L_3(2, 2, 2)$, and we must prove that in fact the type $L_3(2, 2, 2)$ is realized.

We assume toward a contradiction that the triangle type $L_3(2, 2, 2)$ is forbidden.

CLAIM 1. *The configuration $L_4(1, 2, 1; 1, 1; 2)$ is forbidden.*

The following diagram shows that the configuration $C_4(1, 1, 1, 1; 2, 2)$ is realized.

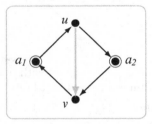

Now consider the following amalgamation diagram.

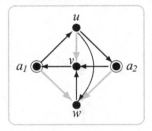

This has no completion in Γ, and the factors are of the form $C_4(1, 1, 1, 1; 2, 2)$ and $L_4(1, 2, 1; 1, 1; 2)$. The claim follows.

CLAIM 2. *The configuration $IC_3(121; 112)$ is forbidden.*

We show first that the configuration $L_4(1, 1, 1; 2, 2; 1)$ is realized.
We consider the following diagram.

Any completion has a 2-arc between vertex 1 and 2, and $2 \xrightarrow{1} 1$ would give $L_3(1, 1, 1; 1, 1; 2)$, contradicting the previous claim. So the completion has $1 \xrightarrow{1} 2$ and thus the configuration $L_4(1, 1, 1; 2, 2; 1)$ is realized.

Now consider the following diagram.

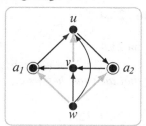

This has no completion in Γ and the factors are of the form $L_4(1, 1, 1; 2, 2; 1)$ and $IC_3(1, 2, 1; 112)$. The claim follows.

CLAIM 3. *The configuration $L_4(1, 2, 1; 1, 2; 2)$ is forbidden.*

We show first that the configuration $C_3I(2, 2, 1; 2, 1, 1)$ is realized. This is the unique completion of the following diagram.

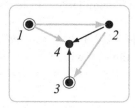

Now consider the following diagram.

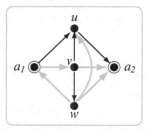

This has no completion in Γ. One factor has the form $C_3I(2, 2, 1; 2, 1, 1)$ and the other has the form $L_4(1, 2, 1; 1, 2; 2)$.

After these preparations we reach a contradiction as follows. We consider the following diagram.

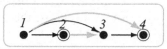

This has three possible completions. Each contradicts one of our three claims.

This contradiction completes the proof of the proposition.   □

22.5. Generalizing Lemma 21.4.25

First we generalize Lemma 21.4.23.

LEMMA 22.5.1. *Suppose that* Γ *is a homogeneous primitive 3-constrained 2-multi-tournament. Suppose that the triangle type* $C_3(1, 1, 1)$ *is forbidden and the triangle types*

$$L_3(1, 1, 2), \ L_3(2, 2, 1),$$
$$C_3(1, 1, 2), \ C_3(2, 2, 1), \ C_3(2, 2, 2)$$

are realized. Then the triangle types $L_3(1, 2, 1)$ *and* $L_3(2, 1, 1)$ *are also realized.*

PROOF. The hypotheses are preserved under reversal of 1-arcs. Therefore it suffices to deal with the triangle type $L_3(1, 2, 1)$. So we suppose toward a contradiction that the triangle types $C_3(1, 1, 1)$, $L_3(1, 2, 1)$ are both forbidden.

Then we still have a degree of symmetry: our hypotheses are preserved under reversal of 2-arcs.

CLAIM 1. *The triangle type* $L_3(2, 1, 1)$ *is realized.*

Consider any diagram of the following form, with orientation of the 2-arcs specified.

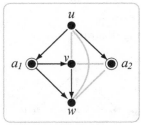

This has no completion in Γ.

On the other hand, if the appropriate triangles embed into Γ, then the first factor (omitting a_2) is the unique completion of the following diagram.

In such cases it follows that the second factor is forbidden.

Then if the appropriate triangles embed into Γ this will force the following diagram to have a unique solution, with the 2-arc between u and v given the opposite orientation to that specified in the factor.

The relevant triangles for this argument are the four triangles (a_1, u, w), (u, v, w), (a_2, u, w), and (a_2, v, w). If we choose the orientations

$$w \xrightarrow{2} u \xrightarrow{2} v, \qquad\qquad u \xrightarrow{2} a_2 \xrightarrow{2} w,$$

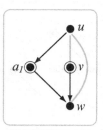

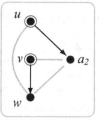

then these triangles have type $L_3(1, 1, 2)$, $C_3(2, 2, 1)$, $C_3(2, 2, 1)$, $L_3(2, 2, 1)$ respectively, and thus are realized in Γ. Thus the following embeds in Γ.

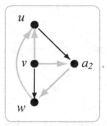

This includes the triangle types $L_3(2, 1, 2)$ and $L_3(1, 2, 2)$. Thus our initial analysis applies to all forms of the amalgamation diagram, and the first factor always embeds in Γ, with the second factor forbidden, regardless of the orientation of the 2-arcs.

But as we have just shown, some form of the second factor must embed in Γ. This proves the claim.

CLAIM 2. *Either the configurations*

$$C_4(1, 1, 2, 2; 2, 2), \; C_4(2, 1, 1, 2; 2, 2)$$

are both forbidden, or the configurations

$$C_4(1, 2, 2, 2; 2, 1), \; C_4(2, 2, 1, 2; 1, 2)$$

are both forbidden.

Considering the following diagram, in which the orientation of the 2-arc between a_1 and w may be chosen in either way.

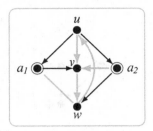

This has no completion in Γ and hence one of the factors must be forbidden. The second factor is $C_4(1, 2, 2, 2; 2, 1)$ and the first factor may be $C_4(1, 1, 2, 2; 2, 2)$ or $C_4(2, 1, 1, 2; 2, 2)$.

Under reversal of 2-arcs the two possibilities for the first factor are interchanged and the second factor is transformed into the configuration $C_4(2, 1, 1, 2; 2, 2)$. Thus either both forms of the first factor or both forms of the second factor must be forbidden, and this is our claim.

CLAIM 3. *The configuration $C_4(2, 1, 1, 2; 1, 2)$ is forbidden.*

We consider the following diagram, which has no completion in Γ.

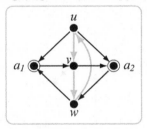

Both factors of this diagram are of the form $C_4(2, 1, 1, 2; 1, 2)$. So the claim follows.

Now we may reach a contradiction. We consider the configuration

$$C_4(2, 1, 1, 2; 1, 2)$$

in two ways as an amalgamation problem.

Namely, we may view this diagram either as the amalgam of the two triangles $(1, 3, 4)$ and $(2, 3, 4)$, or the amalgam of the two triangles $(1, 2, 4)$ and $(1, 3, 4)$. These triangles are of types $C_3(2, 2, 2)$, $L_3(1, 1, 2)$, or $C_3(1, 1, 2)$, all of which are realized in Γ. With the configuration $C_4(2, 1, 1, 2; 1, 2)$ excluded, in the first case the possible completions have $1 \overset{2}{\text{---}} 3$ (in either orientation) and in the second case they have $2 \overset{2}{=} 3$, in either orientation. Thus one of each of the following pairs of configurations must be realized.

$$C_4(2,1,1,2;2,2),\ C_4(1,1,2,2;2,2);$$
$$C_4(2,2,1,2;1,2),\ C_4(1,2,2,2;2,1).$$

But this contradicts our second claim, and this contradiction proves the lemma. □

LEMMA 22.5.2. *Suppose that Γ is a homogeneous 3-constrained 2-multi-tournament for which the triangle type $C_3(1,1,1)$ is forbidden and the triangle types*

$$L_3(1,1,2),\ L_3(2,1,1),\ L_3(2,2,1),$$
$$C_3(1,1,2),\ C_3(2,2,1)$$

are realized. Then the triangle type $L_3(2,2,2)$ is realized.

PROOF. We suppose toward a contradiction that triangle types $C_3(1,1,1)$ and $L_3(2,2,2)$ are both forbidden, and the specified triangle types are realized. Our hypotheses are preserved by reversal of 2-arcs. We will refer to the 2-multi-tournament obtained from a given one T by reversing its 2-arcs as its 2-*reversal*, and denote this by T'.

In the following proof we make repeated reference to the following configurations, which will be referred to by the abbreviated labels indicated.

Label	Structure	Label	Structure
L1	$L_4(1,1,1;2,2;1)$	C1	$C_4(1,1,1,1;2,2)$
L2	$L_4(1,1,1;2,2;2)$	C2	$C_4(1,1,2,1;2,2)$
L3	$L_4(1,1,2;2,2;1)$	C3	$C_4(1,2,2,2;1,1)$
L4	$L_4(1,2,1;1,1;2)$	C4	$C_4(2,2,2,2;1,1)$
L5	$L_4(1,2,1;1,2;2)$		
L6	$L_4(1,2,1;2,1;2)$	D1	$IC_3(1,2,1;1,1,2)$
L7	$L_4(1,2,2;1,1;2)$	D2	$C_3I(1,1,2;2,2,1)$

In some cases their 2-reversals are relevant, so we list these as well, for reference.

Label	Structure
L1′	$C_4(2,1,2,1;1,1)$
L2′	$C_4(1,2,1,1;2,2)$
L7′	$C_4(2,1,2,1;2,1)$

Now we begin our analysis, leading ultimately to a contradiction. Our first three claims will take us to the conclusion that one of $D1$, $C3$, or $L5$ is realized, and this will serve as our real point of departure.

CLAIM 1. *$C1$ and $C4$ are realized.*

For $C1$ use the following amalgamation

and for $C4$ use the same construction with types $\xrightarrow{1}$, $\xrightarrow{2}$ interchanged. In both cases one must check that the triangles occurring as factors are among those assumed realized in Γ; the same applies to a number of similar constructions below (when the factors are in fact triangles).

CLAIM 2. *$L4$ is forbidden.*

The following amalgamation diagram has no completion and the factors are $L4$ and $C1$.

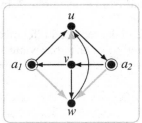

As $C1$ is realized, $L4$ must be forbidden.

CLAIM 3. *One of $D1$, C_3', or $L5$ is realized.*

We consider the following amalgamation problem.

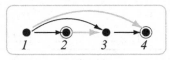

Its completions are $L4$, $D1$, $C3'$, and $L5$. As $L4$ is forbidden the claim follows.

The main line of analysis emerges in the case in which $D1$ is realized. If we make this assumption, we lose the symmetry under 2-reversal, but practically speaking the next lemma restores that symmetry.

CLAIM 4. *If $D1$ is realized then $L1$ and $L1'$ are forbidden.*

The following diagrams have no completions in Γ.

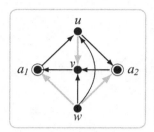 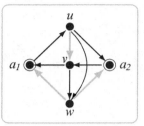

They have the factors $D1$, $L1$ and $L1'$, $D1$ respectively. The claim follows.

CLAIM 5. *If $L1$ is forbidden then $L3$ or $L7$ is realized.*

The following diagram has completions $L1$, $L4$, $L7$, $L3$, and the first two of these are forbidden.

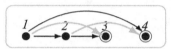

CLAIM 6. *If $L3$ is realized then $D2$ is forbidden. If $L7$ is realized then $C2$ is forbidden.*

We consider the following diagram in two forms, with either orientation of the 2-arc between u and w.

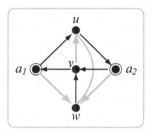

If $u \xrightarrow{2} w$ the factors are $L3$ and $D2$. If $w \xrightarrow{2} u$ the factors are $L7$ and $C2$.

CLAIM 7. *If $C2$ is forbidden then $D2$ is forbidden.*

First, if $C2$ is forbidden then as $L_3(222)$ is forbidden the following diagram gives $C_4(2121; 12)$.

Then if $D2$ is realized the following diagram gives a contradiction.

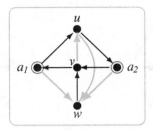

CLAIM 8. *If D2 is forbidden then L7 is realized and C2 is forbidden.*

If *D*2 is forbidden then the following diagram gives *L*7.

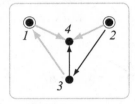

Then Claim 6 applies.

In a certain sense the next claim tends to break the symmetry (and leads eventually to a contradiction).

CLAIM 9. *If C2 is forbidden then L7' is forbidden.*

If the following diagram has a completion, then the structure induced on vertices 1, 2, 3, 4 will be *C*2.

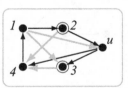

The factors are *C*1 and *L*7'. As *C*1 is realized, *L*7' must be forbidden.

CLAIM 10. *At least one of L1, L1' is realized. In particular D1 is forbidden.*

Suppose that both *L*1 and *L*1' are forbidden. Then we still have symmetry with respect to 2-reversals.

By Claims 5 and 6 we find that *C*2 or *D*2 is forbidden. By Claims 7 and 8 we find that both are forbidden, and *L*7 is realized. By symmetry we also have *L*7' realized. But this contradicts Claim 9.

Thus at least one of *L*1, *L*1' is realized, and the second point follows by Claim 4.

Now we show in two steps that *L*2 is realized.

CLAIM 11. *If L2 is forbidden then L6 is realized and L2' is forbidden.*

The following diagram has completions *L*2, *L*6.

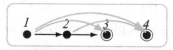

So if $L2$ is forbidden then $L6$ is realized.

The next diagram has no completion and has factors $L6$, $L2'$.

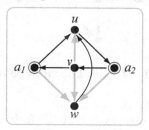

The claim follows.

CLAIM 12. *$L2$ is realized.*

Assuming the contrary, the following diagram has no completion.

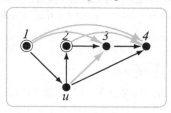

The factors are $L6$ and $L1$. By Claim 11 $L6$ is realized and thus $L1$ is forbidden. But by Claim 6 $L2'$ is also forbidden and applying symmetry we find $L1'$ is forbidden. This contradicts Claim 10.

CLAIM 13. *$L5$ is forbidden and $C3$ is realized.*

The following diagram has no completion.

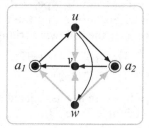

Its factors are $L2'$ and $L5$. By the previous claim and symmetry, $L2'$ is realized. So $L5$ is forbidden.

But by Claim 3 at least one of $D1$, $C3'$, or $L5$ is realized, and by Claim 10 $D1$ is also forbidden. So $C3'$ is realized. But then by symmetry $C3$ is realized.

CLAIM 14. *L3 is realized.*

The following amalgamation forces a realization of *L*3.

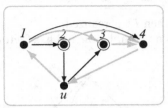

The factors are *C*3 and *C*4, both known to be realized. This proves the claim.

Now we may reach a final contradiction.

By Claims 6 and 8, *L*7 is realized and *C*2 is forbidden. By symmetry *L*7′ is also realized. This contradicts Claim 9.

This contradiction proves the lemma.   □

Now we generalize Lemma 21.4.25.

LEMMA 22.5.3. *Suppose that* Γ *is a homogeneous primitive 2-multi-tournament whose forbidden triangles do not define a free amalgamation class. Suppose that the triangle type*

$$C_3(1, 1, 1)$$

is forbidden and the triangle types

$$L_3(1, 1, 2), \ L_3(2, 2, 1), \ C_3(2, 2, 1), \ C_3(2, 2, 2)$$

are realized.

Then the triangle type

$$C_3(1, 1, 2)$$

is forbidden.

PROOF. Suppose on the contrary that the triangle type $C_3(1, 1, 2)$ is realized. Then Lemma 22.5.1 shows that triangle types $L_3(1, 2, 1)$ and $L_3(2, 1, 1)$ are realized, and then Lemma 22.5.2 shows that the triangle type $L_3(2, 2, 2)$ is realized. So at this point we have the following triangle types realized.

$$L_3(1, 1, 2), \ L_3(1, 2, 1), \ L_3(2, 1, 1), \ L_3(2, 2, 1), \ L_3(2, 2, 2),$$
$$C_3(1, 1, 2), \ C_3(2, 2, 1), \ C_3(2, 2, 2).$$

As the triangles forbidden by Γ are not associated with a free amalgamation class, at least one of the triangle types $L_3(1, 2, 2)$ or $L_3(2, 1, 2)$ must be forbidden.

Our hypotheses are invariant under reversal of both 2-types. So by symmetry we may suppose that

$$L_3(1, 2, 2) \text{ is forbidden.}$$

In the course of our argument we will focus on the following configurations of order 4, and eventually bring in two more toward the end of the argument. In some cases we also work with the 1-reversal of the configuration. For a 2-multi-tournament A, the 1-reversal will be denoted by A'. Note that our current hypotheses are invariant under reversal of $\xrightarrow{1}$.

I.D.	Structure	I.D.	Structure	1-reversal
$L1$	$L_4(221; 11; 1)$	$C1$	$C_4(1111; 22)$	
$L2$	$L_4(221; 11; 2)$	$C2$	$C_4(1112; 12)$	
		$C3$	$C_4(1112; 21)$	$IC_3(112; 112)$
		$C4$	$C_4(1112; 22)$	
		$C5$	$C_4(2112; 11)$	
		$C6$	$C_4(2212; 11)$	$IC_3(211; 221)$

CLAIM 1. *The configuration $C1$ is realized.*

As previously, the following diagram, involving triangles of types $L_3(1, 1, 2)$ and $C_3(1, 1, 2)$, suffices.

CLAIM 2. *The configuration $C4$ is realized iff the configuration $C6$ is realized.*

The following diagram has factors $C6$ and $C1$, and there is a unique completion, with $a_1 \xrightarrow{2} a_2$.

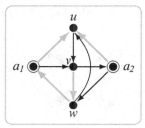

This completion contains a copy of $C4$. Thus

$$C6 \Longrightarrow C4.$$

If we consider the same diagram with the 1-arcs reversed in the 4-cycle $(u, 2, 3, 4)$, then the factors are $C4$ and $C1$, and in a completion we again have

$1 \xrightarrow{2} 3$, and the configuration induced on $(4, 1, 3, 2)$ is $C6$, so

$$C4 \Longrightarrow C6.$$

The claim follows.

CLAIM 3. *If the configuration $C4$ is forbidden then the configuration $C3$ is realized.*

We suppose the configuration $C4$ is forbidden.

The following diagram has as its factors triangles of type $L_3(1, 1, 2)$ and $C_3(1, 1, 2)$.

As $C4$ is forbidden the completion does not have $1 \xrightarrow{2} 3$. Therefore the completion has $1 \xrightarrow{1} 3$, and is $C2$.

Now we consider the following amalgamation diagram, with factors $C5$ and $C2$.

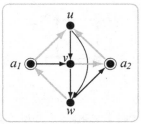

As this has no completion, we conclude that $C5$ is forbidden.

Finally, consider the following diagram, with factors triangles of type $C_3(1, 1, 2)$ and $C_3(2, 2, 1)$.

In the completion since $C5$ is forbidden we have $3 \xrightarrow{1} 2$ or $2 \xrightarrow{2} 3$ and the result is then $C3$ or $C6$ respectively. But as $C4$ is forbidden, $C6$ is also forbidden, and thus $C3$ is realized.

The claim follows.

Now we turn to a close consideration of the configuration $L2$.

CLAIM 4. *If the configuration C6 is forbidden then the configuration L2 is forbidden.*

We first consider the following diagram, with factors $C3$ and $C3'$, and no completion.

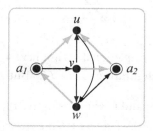

Thus at least one of the factors $C3$ or $C3'$ is forbidden.

As $C6$ is forbidden, also $C4$ is forbidden and $C3$ is realized. Then $C3'$ is forbidden.

But then by the dual of Claim 3 under reversal of $\xrightarrow{1}$, the configuration $C4'$ is realized; and then by the dual of Claim 2, the configuration $C6'$ is realized.

The following diagram has factors $C6'$ and $L2$, and no completion.

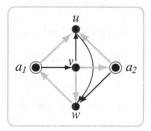

Thus L_2 is forbidden.

CLAIM 5. *If C6 is forbidden then L2 is realized.*

We suppose $C6$ is forbidden, so by Claims 2 and 3 $C3$ is realized. The following diagram has factors $C3$ and $L1$, and no completion.

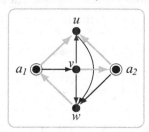

Thus $L1$ is forbidden.

The following diagram has factors triangles of type $L_3(2, 1, 1)$ and $L_3(2, 2, 1)$.

As $C6$ is forbidden the completion does not have $4 \xrightarrow{2} 1$. Thus it has $1 \xrightarrow{1} 4$ or $1 \xrightarrow{2} 4$, giving the configuration $L1$ or $L2$. But L_1 is forbidden and thus L_2 is realized.

The previous two claims and Claim 2 now give the following.

CLAIM 6. *The configurations $C4$ and $C6$ are realized.*

We now work with two more configurations.

I.D.	Structure	1-reversal
$C7$	$C_4(2112; 12)$	$C_4(1122; 21)$
$C8$	$C_4(2112; 22)$	$C_4(2212; 12)$

CLAIM 7. *The configuration $C7$ is forbidden.*

The following diagram has factors $C7$ and $C4$ and no completion.

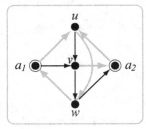

The claim follows.

CLAIM 8. *The configurations $C8$ and $C8'$ are realized.*

The following diagram has factors triangles of type $C_3(1, 1, 2)$ and $C_3(2, 2, 2)$.

The completion may have $2 \xrightarrow{1} 3$, $2 \xrightarrow{2} 3$, or $3 \xrightarrow{1} 2$ and thus the completion may be $C7$, $C8$, or $C7'$ respectively. But $C7$ is forbidden, and by symmetry

$C7'$ is also forbidden. So $C8$ is realized, and by symmetry $C8'$ is also realized.

This proves the claim, and now to reach a contradiction it suffices to consider the following diagram, with factors $C8$ and $C8'$.

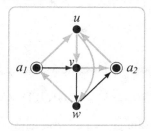

This contradiction proves the lemma.   □

22.6. Around entry #8

PROPOSITION 22.6.1. *Suppose that* Γ *is an infinite homogeneous primitive 2-multi-tournament for which the triangle types*

$$C_3(1,1,1), \ C_3(1,1,2)$$

are forbidden, the triangle types

$$L_3(1,1,1), \ L_3(2,2,1), \ C_3(2,2,1)$$

are realized, and there is no nontrivial $\emptyset$-definable partial order.

Then all other triangle types are realized, as in entry #8 of our table.

PROOF. By Lemma 21.4.22, Proposition 22.3.1, and the hypothesis the following triangle types are realized.

$$L_3(1,1,1), \ L_3(1,1,2), \ L_3(1,2,1), \ L_3(2,1,1), \ L_3(2,2,1),$$
$$C_3(2,2,1), \ C_3(2,2,2).$$

That is, Lemma 21.4.22 and the hypotheses cover everything listed other than triangle type $C_3(2,2,2)$, and if triangle type $C_3(2,2,2)$ were omitted then Proposition 22.3.1 would give a contradiction.

This leaves the following types to be considered.

$$L_3(1,2,2), \ L_3(2,1,2), \ L_3(2,2,2).$$

CLAIM 1. *Triangle types* $L_3(1,2,2)$ *and* $L_3(2,1,2)$ *are realized.*

As our hypotheses are closed under reversal of both 2-types, it suffices to treat the case of $L_3(2,1,2)$. So suppose toward a contradiction that the triangle type $L_3(2,1,2)$ is also forbidden. If the type $L_3(1,2,2)$ is realized then the following diagram gives a contradiction.

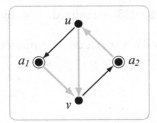

So at this point we suppose we have the following forbidden triangle types.

$$L_3(1, 2, 2), \ L_3(2, 1, 2), \ C_3(1, 1, 1), \ C_3(1, 1, 2).$$

Then the following diagram gives a contradiction.

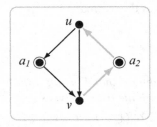

CLAIM 2. *Triangle type $L_3(2, 2, 2)$ is realized.*

From the following diagram we see that $C_4(1, 1, 1, 2; 2, 2)$ is realized.

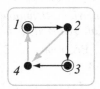

A similar diagram forces $C_4(2, 2, 2, 2; 1, 1)$.
Then the following diagram forces $L_3(2, 2, 2)$.

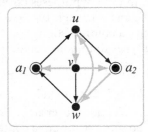

These two claims prove the proposition.   □

22.7. Around entry #9

Now we aim at the following.

PROPOSITION 22.7.1. *Suppose that Γ is a homogeneous primitive 2-multi-tournament with no $\emptyset$-definable partial order, whose forbidden triangles do not define a free amalgamation class, in which the triangle types*

$$C_3(1,1,1),\ C_3(2,2,2)$$

are realized and the triangle type $C_3(1,1,2)$ is forbidden.

Then all other triangle types are realized except possibly $L_3(2,2,1)$ and $L_3(1,1,1)$.

LEMMA 22.7.2. *Suppose that Γ is a homogeneous primitive 2-multi-tournament with no $\emptyset$-definable partial order, whose forbidden triangles do not define a free amalgamation class. Suppose that the triangle types $C_3(1,1,1)$ and $C_3(2,2,2)$ are realized and $L_3(1,2,1)$ is forbidden.*

Then the triangle type

$$L_3(2,1,1)$$

is forbidden.

PROOF. We suppose the contrary and analyze the situation at length. By Lemma 21.4.27 we then have the following conditions satisfied.

Forbidden: $L_3(1,2,1)$.

Realized: $L_3(1,1,1), L_3(1,1,2), L_3(2,1,1),$
$C_3(1,1,1), C_3(1,1,2), C_3(2,2,2).$

We observe that our conditions are preserved by reversal of 2-arcs. We denote the 2-reversal of a 2-multi-tournament A by A'.

In what follows, we refer to the following specific configurations of order 4.

I.D.	Structure
L1	$L4(1,1,1;1,1;2)$
L2	$L_4(1,1,1;1,2;2)$
C1	$C_4(1,1,1,1;1,1)$
C2	$C_4(1,1,1,1;1,2)$
C3	$C_4(1,1,1,1;2,2)$
C4	$C_4(1,2,1,1;2,1)$

CLAIM 1. *Triangle types $L_3(2,2,1)$ and $L_3(2,1,2)$ are realized.*

The following diagram has factors of type $L_3(2,1,1)$ and $L_3(1,1,2)$, and forces triangles of types $L_3(2,2,1)$ and $L_3(2,1,2)$ to be realized.

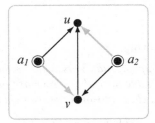

CLAIM 2. *The triangle type $L_3(1,2,2)$ and the configurations $L1$ and $C1$ are realized.*

The argument proceeds in a similar fashion in all cases. We fix a point $a \in \Gamma$ and set $\Gamma^1 = a^1$, $\Gamma^2 = a^p$ where the type p is $\xleftarrow{1}$ when the target is $L_3(2,2,1)$ or $C1$, and is $\xrightarrow{2}$ when the target is L_1.

We first discuss the case of $C1$ $(C_4(1,1,1,1;1,1))$. In this case, we seek $v_1, v_2 \in \Gamma_1$ and $u \in \Gamma_2$ with $v_1, v_2 \xrightarrow{1} u$.

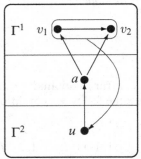

If this fails then we have a partial function $f : \Gamma^2 \to \Gamma_1$ defined by

$$f(u) \xrightarrow{1} u.$$

Since the triangle $C_3(1,1,1)$ is realized, this function is in fact total.

Let $\sim$ be the equivalence relation on Γ^2 defined by $f(x) = f(y)$. Then f gives a bijection between $\Gamma^2/\sim$ and Γ^1. It follows easily that $\sim$ is a congruence and for u_1, u_2 in distinct $\sim$-classes, the type of (u_1, u_2) determines the type of $(u_1, f(u_2))$. But there are two such 2-types in Γ^2 and three realized between u and elements of Γ^1 other than $f(u)$, a contradiction.

Thus we have the configuration $C1$.

We argue similarly to get the configuration $L_3(1,2,2)$, now looking for $v_1, v_2 \xrightarrow{2} u$. If this is not realized then we define $f : \Gamma^2 \to \Gamma^1$ by by

$$f(u) \xrightarrow{2} u$$

and reach the same contradiction as above, taking into account the existence of a triangle of type $C_3(1,1,2)$.

For the configuration $L1 = L_4(1, 1, 1; 1, 1; 2)$ we proceed similarly. Now in our diagram we have $a \xrightarrow{2} u$. The same argument then gives the configuration $L1$, taking into account the existence of a triangle of type $L_3(1, 1, 2)$.

CLAIM 3. *The configurations $C2$ and $C2'$ are forbidden.*

The following diagram has factors $L1$ and $C2$, and no completion.

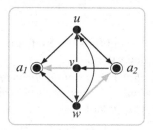

Thus $C2$ is forbidden, and by symmetry the same applies to $C2'$.

CLAIM 4. *The configuration $C3$ is realized.*

The following amalgamation diagram has factors of type $C_3(1, 1, 2)$ and $L_3(1, 1, 2)$.

As the configurations $C2$ and $C2'$ are forbidden, we have $1 \xrightarrow{2} 3$ in the completion. Thus $C3$ is realized.

CLAIM 5. *The configuration $C4$ is forbidden.*

The following amalgamation diagram has factors $C3$, $C4$ and no completion.

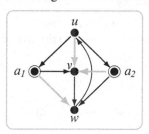

The claim follows.

Now we arrive at a contradiction. The following amalgamation diagram has factors $C1$ and $L2$, which are realized.

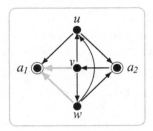

In the completion $a_1 \xrightarrow{1} a_2$ and thus the configuration $C4$ must be realized. This contradiction completes the proof.   □

LEMMA 22.7.3. *Suppose that* Γ *is a homogeneous primitive 2-multi-tournament with no $\emptyset$-definable partial order, whose forbidden triangles do not define a free amalgamation class. Suppose that the triangle types*

$$C_3(1,1,1), \ C_3(2,2,2)$$

are realized and $L_3(1,2,1)$ *is forbidden.*
    Then the triangle types

$$L_3(1,2,2), \ L_3(2,1,2)$$

are forbidden.

PROOF. By symmetry it suffices to deal with $L_3(1,2,2)$.

So suppose toward a contradiction that the triangle type $L_3(1,2,2)$ is realized.

Then by Lemmas 21.4.27, 21.4.29, and 22.7.2 we have the following situation.

Forbidden: $L_3(1,2,1), L_3(2,1,1)$.

Realized: $L_3(1,1,1), L_3(1,1,2), L_3(1,2,2), L_3(2,2,1),$
        $C_3(1,1,1), C_3(1,1,2), C_3(2,2,1), C_3(2,2,2)$.

CLAIM 1. *The configuration* $IC_3(1,1,1;1,1,1)$ *is forbidden and the configuration* $C_4(1,1,1,1;1,1)$ *is realized.*

The following diagram has no completion. Here we do not specify the orientations of the 2-arcs as they make no difference.

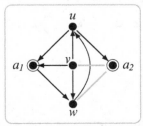

The second factor of this diagram is the unique completion of the following.

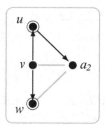

The factors may be taken to be e.g. $L_3(1, 1, 2)$ and $L_3(2, 2, 1)$. So the first factor must be forbidden, and this is $IC_3(1, 1, 1; 1, 1, 1)$.

Now view the first factor as an amalgam of a_1uw with vuw. Then the completion must have $u \xrightarrow{1} v$, and this gives $C_4(1, 1, 1, 1; 1, 1)$.

This proves the claim.

Now we may reach a contradiction. The following diagram has no completion.

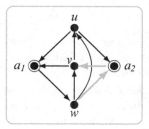

The first factor is $C_4(1, 1, 1, 1; 1, 1)$. The second factor may be viewed as the unique completion of the amalgam of a_2uw and a_2vw, triangles of types $L_3(1, 1, 2)$ and $L_3(2, 2, 1)$. □

LEMMA 22.7.4. *Suppose that Γ is a homogeneous primitive 2-multi-tournament with no $\emptyset$-definable partial order, whose forbidden triangles do not define a free amalgamation class, in which the triangle types*

$$C_3(1, 1, 1), \quad C_3(2, 2, 2)$$

are realized. Then the triangle types

$$L_3(1, 2, 1), \quad L_3(1, 2, 2), \quad L_3(2, 1, 1), \quad L_3(2, 1, 2)$$

are realized.

PROOF. Our hypotheses are invariant under reversal of either 2-type. By symmetry, it suffices to treat the case of $L_3(1, 2, 1)$. Suppose toward a contradiction that the triangle type

$$L_3(1, 2, 1)$$

is forbidden.

By Lemmas 21.4.27, 21.4.29, 22.7.2, and 22.7.3 we then have the following

Forbidden: $L_3(1,2,1)$, $L_3(1,2,2)$, $L_3(2,1,1)$, $L_3(2,1,2)$.

Realized: $L_3(1,1,1)$, $L_3(1,1,2)$, $L_3(2,2,1)$,

$C_3(1,1,1)$, $C_3(1,1,2)$, $C_3(2,2,1)$, $C_3(2,2,2)$.

As $L_3(2,1,2)$ is forbidden, also $L_3(2,2,2)$ is realized. Thus four triangle types are forbidden and the rest are realized.

Now fix $a \in \Gamma$ and let $\Gamma^1 = a^1$, $\Gamma^2 = a^2$. Note that Γ^1 is a transitive $\overset{1}{\longrightarrow}$-tournament.

By our assumptions for $u \in \Gamma^1$ and $v \in \Gamma^2$ the only possible relations are

$$u \overset{1}{\longrightarrow} v; \qquad\qquad v \overset{2}{\longrightarrow} u.$$

If we have $u \in \Gamma^1$ and $v_1, v_2 \in \Gamma^2$ with $u \overset{1}{\longrightarrow} v_1, v_2$ then (a, v_1, v_2) has type $L_3(1,2,1)$, a contradiction. Therefore we may define $f : \Gamma^2 \to \Gamma^1$ by

$$u \overset{1}{\longrightarrow} f(u).$$

Then the relations $f(u) \overset{1}{\longrightarrow} v$ and $f(u) \overset{1}{\longleftarrow} v$ give distinct relations between Γ^2 and Γ^1, but there is only one such relation.

This contradiction proves the lemma.   □

LEMMA 22.7.5. *Suppose that Γ is a homogeneous primitive 2-multi-tournament with no $\emptyset$-definable partial order whose triangle constraints are not associated with a free amalgamation class. Suppose that the triangle types $C_3(1,1,1)$ and $C_3(2,2,2)$ are realized and the triangle type $C_3(1,1,2)$ is forbidden.*

Then the triangle type

$$L_3(1,1,2)$$

is realized.

PROOF. Suppose the contrary.

We consider the following diagram, with the orientation of the arc between v and w to be determined.

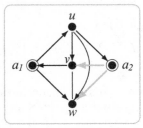

This has no completion. If we consider the first factor as an amalgam of a_1uv with a_1uw then the orientation of the arc between v and w will be determined. We then consider the second factor.

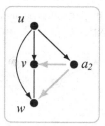

If $v \xrightarrow{1} w$ then this is the unique amalgam of a_2uv with a_2vw. If $w \xrightarrow{1} v$ then this is the unique amalgam of a_2uw with a_2vw.

The triangles involved are of types $L_3(1, 2, 1)$, $L_3(2, 1, 2)$, or $L_3(1, 2, 2)$. These are afforded by Lemma 22.7.4.

Thus we have a contradiction and the proof is complete. $\square$

LEMMA 22.7.6. *Suppose that Γ is a homogeneous primitive 2-multi-tournament with no $\emptyset$-definable partial order. Suppose that the triangle types $C_3(1, 1, 1)$ and $C_3(2, 2, 2)$ are realized and the triangle type $C_3(1, 1, 2)$ is forbidden. Then the triangle type*

$$C_3(2, 2, 1)$$

is realized.

PROOF. Lemmas 22.7.5 and 21.4.34. $\square$

LEMMA 22.7.7. *Suppose that Γ is a homogeneous primitive 2-multi-tournament with no $\emptyset$-definable partial order, whose forbidden triangles do not define a free amalgamation class. Suppose that the triangle types*

$$L_3(2, 2, 1), \quad C_3(1, 1, 1), \quad C_3(2, 2, 2)$$

are realized and the triangle type

$$C_3(1, 1, 2)$$

is forbidden.

Then the triangle type

$$L_3(2, 2, 2)$$

is realized

PROOF. By Lemmas 22.7.4, 22.7.5, and 22.7.6 we have the following.

Forbidden: $C_3(1, 1, 2)$.

Realized: $L_3(1, 1, 2)$, $L_3(1, 2, 1)$, $L_3(1, 2, 2)$, $L_3(2, 1, 1)$, $L_3(2, 1, 2)$,
$L_3(2, 2, 1)$, $C_3(1, 1, 1)$, $C_3(2, 2, 1)$, $C_3(2, 2, 2)$.

Assume toward a contradiction that the triangle type $L_3(2, 2, 2)$ is forbidden.

CLAIM 1. *The configuration $IC_3(1, 1, 2; 2, 2, 2)$ is realized.*

This is forced by the following amalgamation, with factors $C_3(2,2,2)$ and $L_3(2,2,1)$, since the triangle types $C_3(1,1,2)$ and $L_3(2,2,2)$ are forbidden.

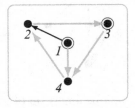

CLAIM 2. *The configuration $C_4(2,2,2,2;1,1)$ is realized.*

This is forced by the following amalgamation, with factors $C_3(2,2,1)$ and $L_3(2,1,1)$, since the triangle type $L_3(2,2,2)$ is forbidden.

Now we reach a contradiction. The following amalgamation has factors $IC_3(1,1,2;2,2,2)$ and $C_4(2,2,2,2;1,1)$, and no completion.

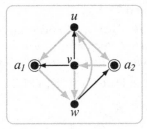

This completes the proof.   □

LEMMA 22.7.8. *Suppose that Γ is a homogeneous primitive 2-multi-tournament with no $\emptyset$-definable partial order, whose forbidden triangles do not define a free amalgamation class, in which the triangle types*

$$C_3(1,1,1),\ C_3(2,2,2)$$

are realized and the triangle type $C_3(1,1,2)$ is forbidden.

Then the triangle type $L_3(2,2,2)$ is realized.

PROOF. We dealt in the previous lemma with the case in which the triangle type $L_3(2,2,1)$ is realized, so now we suppose that type is forbidden. Assuming toward a contradiction that

Triangle types $L_3(2,2,1)$, $L_3(2,2,2)$, $C_3(1,1,2)$ are forbidden.

Triangle types $C_3(1,1,1)$, $C_3(2,2,2)$ are realized.

Then as the triangle type $L_3(2, 2, 2)$ is forbidden the type $L_3(1, 1, 1)$ is realized, and all other triangle types are realized as well, by Lemmas 22.7.4, 22.7.5, and 22.7.6.

We work here with the following configurations and their reversals, which we denote by $'$.

I.D.	Structure	I.D.	Structure
L1	$L_4(1, 1, 2; 2, 2; 2)$	C1	$C_4(1, 2, 2, 2; 1, 2)$
L2	$L_4(1, 2, 2; 1, 2; 1)$	C2	$C_4(1, 1, 2, 2; 2, 2)$
L3	$L_4(1, 2, 2; 1, 2; 2)$	D1	$IC_3(1, 1, 1; 2, 2, 2)$
L4	$L_4(2, 1, 2; 2, 2; 2)$	D2	$IC_3(2, 2, 2; 2, 2, 1)$

CLAIM 1. *The configuration C2 is realized.*

We use the following amalgamation with a unique solution.

CLAIM 2. *The configuration D1 is forbidden.*

We use the following amalgamation, with factors $D1$ and $C2$.

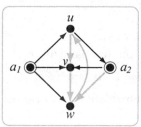

As this has no completion the claim follows.

CLAIM 3. *The configuration L2 is realized.*

We use the following amalgamation.

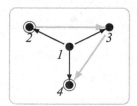

Since $D1$ is forbidden we have $2 \xrightarrow{2} 4$ in the completion, and the claim follows.

CLAIM 4. *The configuration $D2$ is forbidden.*

We use the following amalgamation, with factors $L2$ and $D2$.

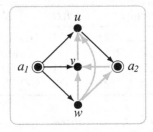

Since there is no completion, the claim follows.

CLAIM 5. *The configuration $L1$ is forbidden.*

We use the following amalgamation, with factors $L1$ and $L2$.

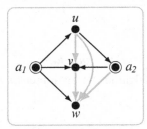

Since there is no completion, the claim follows.

CLAIM 6. *The configuration $L4$ is forbidden.*

We use the following amalgamation, with factors $L2$ and $L4$.

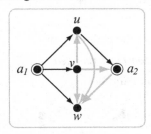

Since there is no completion, the claim follows.

CLAIM 7. *The configuration $L3$ is realized.*

Consider the following amalgamation.

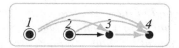

In the completion we must have $1 \xrightarrow{1} 2$, $2 \xrightarrow{1} 1$, or $1 \xrightarrow{2} 2$, which gives one of the configurations $L1$, $L3$, or $L4$, respectively. The claim follows.

CLAIM 8. *The configuration $C1$ is realized.*

We use the following amalgamation.

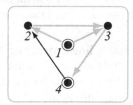

Since the configuration $D2$ is forbidden, the completion has $4 \xrightarrow{1} 1$, which gives $C1$.

Now we reach a contradiction. We know $C1$ is realized and by symmetry also the reversal $C1'$ is realized.

But the following amalgamation has factors $L3$ and $C1'$, and has no completion.

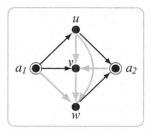

This contradiction completes the proof. □

Combining these results, Proposition 22.7.1 follows.

PROOF OF PROPOSITION 22.7.1. Lemmas 22.7.4, 22.7.5, 22.7.6, and 22.7.8. □

22.8. Forbidden triangles: analysis so far, and what remains

We now have everything in place to prove Proposition 22.1.1. We repeat the statement.

PROPOSITION 22.1.1. *Let Γ be an infinite, primitive, homogeneous 2-multi-tournament not associated with a free amalgamation class. If Γ has a $\emptyset$-definable* *linear order then it is found in the classification in Part I. If not, then either the set of forbidden triangles in Γ defines one of the known 3-constrained 2-multi-tournaments, or one of the following four cases applies.*

1. *Triangle types $C_3(1, 1, 1)$ and $C_3(2, 2, 2)$ are forbidden and all other triangle types are realized.*
2. *Triangle types $C_3(1, 1, 1)$ and $L_3(2, 2, 1)$ are forbidden and all other triangle types are realized.*
3. *Triangle type $C_3(1, 1, 2)$ is forbidden and all other triangle types realized.*
4. *Triangle types $C_3(1, 1, 2)$ and $L_3(2, 2, 1)$ are forbidden and all other triangle types realized.*

PROOF. If there is a $\emptyset$-definable linear order on Γ then Part I of Volume I completes prior work to give the full classification.

If there is no $\emptyset$-definable linear order but there is some $\emptyset$-definable partial order, then we may assume that the relation $\xrightarrow{1}$ is transitive, and Proposition 22.2.1 applies. Thus, up to a change of language, the triangle constraints are as in entry #6 or #7 of Table 21.1.

Now suppose that *there is no $\emptyset$-definable linear order on Γ*. We consider which of the triangle types $C_3(1, 1, 1)$, $C_3(2, 2, 2)$ is realized.

Case 1. Both triangle types $C_3(1, 1, 1)$ and $C_3(2, 2, 2)$ are forbidden in Γ.

Then by Proposition 22.3.1, either the same triangles are forbidden as in $\widetilde{\mathbb{S}(3)}$, up to a change of language, or else we have exceptional case (1): *triangle types $C_3(1, 1, 1)$ and $C_3(2, 2, 2)$ are forbidden and all other triangle types are realized.*

Case 2. Exactly one of the triangle types $C_3(1, 1, 1)$, $C_3(2, 2, 2)$ is forbidden.

We assume

$$C_3(1, 1, 1) \text{ is forbidden; and}$$
$$C_3(2, 2, 2) \text{ is realized.}$$

Then we must consider a number of subcases.

(2a) If the triangle type $L_3(2, 2, 1)$ is forbidden: then by Proposition 22.4.1 either Γ is isomorphic to $\mathbb{S}(4)$ or we have exceptional case (2):
    Triangle types $C_3(1, 1, 1)$ and $L_3(2, 2, 1)$ are forbidden and all other triangle types are realized.
    The case in which the triangle type $C_3(2, 2, 1)$ is forbidden is the same, up to the choice of language.

(2b) Now suppose the triangle types $L_3(2, 2, 1)$ and $C_3(2, 2, 1)$ are both realized. Then Lemma 22.5.3 implies that at least one of the triangle types $L_3(1, 1, 2)$ or $C_3(1, 1, 2)$ is forbidden.
    Suppose first that $C_3(1, 1, 2)$ is forbidden. Then one must consider further whether $L_3(1, 1, 1)$ is realized.

- If $L_3(1, 1, 1)$ is forbidden then Proposition 21.4.19 shows that the remaining triangle types are realized, as in entry #10.
- If $L_3(1, 1, 1)$ is realized then Proposition 22.6.1 shows that we have the same forbidden triangles as in entry #8.

  If on the other hand $L_3(1, 1, 2)$ is forbidden, then we have the same situation, up to a change of language.

Case 3. Both $C_3(1, 1, 1)$ and $C_3(2, 2, 2)$ are realized.

Then Lemma 22.7.4, p. 207 shows that triangle types

$$L_3(1, 2, 1), \ L_3(1, 2, 2), \ L_3(2, 1, 1), \ L_3(2, 1, 2)$$

are realized.

As Γ is not associated with a free amalgamation class, and cannot forbid both $L_3(1, 1, 1)$ and $L_3(2, 2, 2)$, it follows that up to a change of language Γ may be supposed to forbid the triangle type $C_3(1, 1, 2)$.

By Proposition 22.7.1 all other triangle types are realized with the possible exceptions of $L_3(2, 2, 1)$ and $L_3(1, 1, 1)$.

If the triangle type $L_3(2, 2, 1)$ is also realized then the triangle constraints are as in the 3-constrained entry #9, with the possible exception of $L_3(1, 1, 1)$, which should be forbidden as well. If this is not the case then we have the exceptional case (3): *the triangle type $C_3(1, 1, 2)$ is forbidden and all other triangle types realized.*

There remains the case in which $C_3(1, 1, 2)$ and $L_3(2, 1, 1)$ are both forbidden and all other triangle types are realized with the possible exception of $L_3(1, 1, 1)$. But if $L_3(2, 1, 1)$ is forbidden then easily $L_3(1, 1, 1)$ is realized.

So this gives us the fourth and last exceptional case: *the triangle types $C_3(1, 1, 2)$ and $L_3(2, 2, 1)$ are forbidden and all other triangle types realized.*

This concludes the analysis. □

The four exceptional cases remaining are those whose treatment, under the assumption of 3-constraint, required consideration of amalgamation diagrams of order six.

In the cases where amalgamation diagrams of order five were sufficient, we concluded the argument by a careful consideration of the possible structures of order four occurring in Γ. The remaining four cases are more demanding, but one anticipates that arguments of the same kind would be adequate. Broadly speaking, the question divides roughly into two parts:

(I) Show that any configuration of order 4 which does not have a forbidden triangle is itself realized (4-triviality);

(II) Reach a contradiction somewhat as in the cases just treated by considering which factors of order 5 are realized in Γ, rather than factors of order 4.

Having brought the analysis of triangle constraints down to this point, we leave it there. In our final section we review where this leaves us, relative to the general classification problem for homogeneous 2-multi-tournaments.

22.9. 2-multi-tournaments: conclusion

22.9.1. Triangle constraints: an overview. At this stage, we know the *imprimitive* homogeneous 2-multi-tournaments and we have a classification of the primitive 3-*constrained* homogeneous 2-multi-tournaments which can serve as a point of departure for a more general analysis of the pattern of forbidden triangles in an arbitrary primitive infinite homogeneous 2-multi-tournament.

Proposition 22.1.1 brings that analysis down to four delicate cases remaining to be eliminated. Once that has been achieved, according to each possible set of forbidden triangles which is not compatible with some form of free amalgamation, one wants a particular classification theorem. Some of these have already been given in Part I. Another was given in §22.4.

This remains a large subject. But we can now see clearly what needs to be done to reduce the classification problem to that in which the pattern of forbidden triangles is consistent with free amalgamation. This would still not bring us down to the "generic" case, but covers the bulk of the exceptional cases. One must also deal at an early stage with the exceptional 2-multi-tournaments $\mathbb{S} * \mathbb{S}$, $\mathbb{S} * H_n$, and $\widetilde{\mathcal{P}(3)}$, which do not correspond to free amalgamation classes but which typically have no triangle constraints, and at worst forbid monochromatic ones in a single color.

We next review the current catalog of known examples in detail.

22.9.2. Triangle constraints: the catalog. As the catalog of known homogeneous 2-multi-tournaments is extensive, we conclude this chapter by presenting that catalog once more in a completely explicit form organized according to the constraints of order 3 involved.

First, we give a list of the various classes of examples which occur in a natural order, moving from the most constrained to the least constrained types. We will then present these examples in terms of the minimal constraints. For this purpose, it will be convenient to give each example, or class of examples, a brief symbolic name.

We give the list up to a permutation of the language. This list was given less formally in §20.2, sometimes with some variation in the choice of language, and more overlap between categories.

Known homogeneous 2-multi-tournaments.

1. Degenerate type: $\xrightarrow{2}$ is forbidden.
   (i–v) The homogeneous tournaments I, C_3, $\mathbb{Q}$, $\mathbb{S}$, T^∞. Labels: I, C, Q, S, ∞.

2. Imprimitive with equivalence $\overset{1}{\sim}$: forbid $C_3(1, 1, 2)$, $L_3(1, 1, 2)$, $L_3(1, 2, 1)$, $L_3(2, 1, 1)$.
   (a) Composition $T_2[T_1]$ (T_1, T_2 of type C_3, $\mathbb{Q}$, $\mathbb{S}$, or T^∞): 16 cases. Additional constraints derived from T_1 and T_2 separately. Labels $L_2[L_1]$.

(b) Shuffled
   (i) $\mathbb{Q}^{(n)}$, $2 \leq n \leq \infty$ properly shuffled of type $\mathbb{Q}$. Label Q^n.
   (ii) $\mathbb{S}^{(n)}$, $2 \leq n \leq \infty$, properly shuffled of type $\mathbb{S}$. Label S^n.
(c) General type
   (i) Semi-generic (parity constraint) with component of type $T = \mathbb{Q}$, $\mathbb{S}$, T^∞. Label $\infty \dot{*} T$.
   (ii) Generic of type T with $2 \leq n \leq \infty$ classes. Label $n * T$.
3. Primitive with triangle constraints involving both 2-types.
  (a) Finite: pentagram. Label 5.
  (b) $\xrightarrow{1} \cup \xrightarrow{2}$ a linear order: forbid $C_3(1,1,1)$, $C_3(1,1,2)$, $C_3(2,2,1)$, $C_3(2,2,2)$.
   (i) Generic permutation. Label $Q * Q$.
   (ii) Linear extension of generic partial order. Label $P \subseteq Q$.
   (iii) Generically ordered $\mathbb{S}$. Label $Q * S$.
   (iv) Generically ordered Henson or random graph. Label $Q * H_n$ ($n \leq \infty$).
  (c) No definable linear order, $\xrightarrow{1}$ a partial order: forbid $L_3(1,1,2)$, $C_3(1,1,1)$, $C_3(1,1,2)$.
   (i) Non-generically de-symmetrized partial order forbidding $C_3(2,2,1)$. Label $3C_7$ or $\tilde{\mathcal{P}}^-$.
   (ii) Generically de-symmetrized partial order. Label $3C_6$ or $\tilde{\mathcal{P}}$.
  (d) No 2-type transitive; infinite
   (i) Both $C_3(1,1,1)$ and $C_3(2,2,2)$ forbidden.
    De-symmetrized myopic local order.
    Forbid $C_3(1,1,1)$, $C_3(2,2,2)$, $L_3(1,1,1)$, $L_3(1,2,2)$, $L_3(2,1,2)$. Label: $3C_{11}$ or $S(3)$.
   (ii) $C_3(1,1,1)$ forbidden, $C_3(2,2,2)$ realized.
    (A) 4-myopic local order $\mathbb{S}(4)$
     Forbid $C_3(1,1,1)$, $C_3(1,1,2)$, $L_3(1,2,1)$, $L_3(2,1,1)$, $L_3(2,2,1)$, $L_3(2,2,2)$. Label $3C_{12}$ or $S(4)$.
    (B) Forbid $C_3(1,1,1)$, $C_3(1,1,2)$, L^1_{n+1} ($2 \leq n \leq \infty$). Label $3C_{8:n}$. (For $n = 2$: also called $3C_{10}$.)
   (iii) $C_3(1,1,1)$, $C_3(2,2,2)$ realized.
    Forbid $C_3(1,1,2)$, $L_3(1,1,1)$. Label $3C_9$.
4. No non-monochromatic triangles forbidden.
  (a) Free product $\mathbb{S} * \mathbb{S}$, $\mathbb{S} * H_n$. Label $S * S$, $S * H_n$
  (b) De-symmetrized local partial order. Label $\widetilde{\mathcal{P}(3)}$.
  (c) Free amalgamation classes: forbid a family of $\xrightarrow{1}$-tournaments. Label $\tilde{H}_T$.

Remark 22.9.1. $\mathbb{S}$ (the generic local order) and $\mathcal{P}(3)$ (the generic local partial order) are characterized by forbidden structures of order 4. We also have the structures $T * A$ with T a homogeneous tournament, A a homogeneous graph or tournament, and both having strong amalgamation. $A * T^\infty$ is listed as $A * H_n$ with $n = \infty$. This is also the generic 2-splitting of the tournament A.

The 2-multi-tournaments $\widetilde{H_\mathcal{T}}$ are the generic de-symmetrizations of the directed graphs $H_\mathcal{T}$. We have the (doubly) de-symmetrized Henson graph $\widetilde{H}_n$ as a special case $(T^\infty * H_n)$.

Now we give the list again in tabular form in terms of forbidden substructures. When there are constraints of order greater than 3 (required or optional) we discuss this afterward. We omit the degenerate cases. Our conventions for assigning the 2-type labels 1, 2 are as follows.

(a) In imprimitive structures: put $\xrightarrow{1}$ within the equivalence classes, and $\xrightarrow{2}$ between them.

(b) In structures presented as linear extensions of partial orders $(A, \prec, <)$, $\xrightarrow{1}$ is $\prec$ and $\xrightarrow{2}$ is $\perp$ & $<$.

(c) In free products of strong amalgamation classes $K_a * K_b$ with K_a, K_b classes of finite tournaments with edge relations $\xrightarrow{a}$ and $\xrightarrow{b}$ respectively, we set

$$\xrightarrow{1} \text{ is } \xrightarrow{a} \& \xrightarrow{b}, \text{ and } \xrightarrow{2} \text{ is } \xrightarrow{a} \& \xleftarrow{b}.$$

(d) In free products of strong amalgamation classes $K_a * K_b$ with K_a a classes of finite tournaments and K_b a class of finite graphs with edge relations $\xrightarrow{a}$ and $\overset{b}{-}$ respectively, we set

$$\xrightarrow{1} \text{ is } \xrightarrow{a} \& \overset{b}{-}, \text{ and } \xrightarrow{2} \text{ is } \xrightarrow{a} \& \perp^b.$$

(e) In a de-symmetrized directed graph, $\xrightarrow{1}$ is the original arc relation in the digraph and $\xrightarrow{2}$ is the de-symmetrized non-arc relation.

The first two tables below specify the *triangle constraints* in the imprimitive and primitive cases, respectively. The third table describes additional constraints where appropriate. The constraints of order greater than 3 are associated with forms of $\mathbb{S}$, $\mathcal{P}(3)$, imprimitive graphs with a finite number of equivalence classes (at least 3), the parity constraint in the semi-generic case, and free amalgamation classes.

	3-cycles				L_3								More
I.D.	111	112	221	222	111	112	121	122	211	212	221	222	
$C[C]$		✗	✗		✗	✗	✗		✗		✗	✗	
$C[Q]$	✗	✗	✗			✗	✗		✗		✗	✗	
$C[S]$		✗	✗			✗	✗		✗		✗	✗	✓
$C[\infty]$		✗	✗			✗	✗		✗		✗	✗	
$Q[C]$		✗	✗	✗	✗	✗	✗		✗		✗		
$Q[Q]$	✗	✗	✗	✗		✗	✗		✗		✗		
$Q[S]$		✗	✗	✗		✗	✗		✗		✗		✓
$Q[\infty]$		✗	✗	✗		✗	✗		✗		✗		
$S[C]$		✗	✗		✗	✗	✗		✗		✗		✓
$S[Q]$	✗	✗	✗			✗	✗		✗		✗		✓
$S[S]$		✗	✗			✗	✗		✗		✗		✓
$S[\infty]$		✗	✗			✗	✗		✗		✗		✓
$\infty[C]$		✗	✗		✗	✗	✗		✗		✗		
$\infty[Q]$	✗	✗	✗			✗	✗		✗		✗		
$\infty[S]$		✗	✗			✗	✗		✗		✗		✓
$\infty[\infty]$		✗	✗			✗	✗		✗		✗		
Q^n	✗	✗	✗	✗		✗	✗		✗		$n=2$		$2<n<\infty$
S^n		✗		$n=2$		✗	✗		✗		$n=2$		✓
$\infty \hat{*} Q$		✗				✗	✗		✗				✓
$\infty \hat{*} S$		✗				✗	✗		✗				✓
$\infty \hat{*} \infty$		✗				✗	✗		✗				✓
$n * Q$	✗	✗		$n=2$		✗	✗		✗		$n=2$		✓
$n * S$		✗		$n=2$		✗	✗		✗		$n=2$		✓
$n * \infty$		✗		$n=2$		✗	✗		✗		$n=2$		✓

TABLE 22.1. Imprimitive homogeneous 2-multi-tournaments: triangle constraints.

I.D.	Primitive Cases: 3-cycles 111	112	221	222	L_3 111	112	121	122	211	212	221	222	More
5	✗	✗		✗	✗		✗	✗	✗	✗	✗	✗	
$Q*Q$	✗	✗	✗	✗	✗						✗		
$P \subseteq Q$	✗	✗	✗	✗	✗								
$Q*S$	✗	✗	✗	✗									✓
$Q*H_n$	✗	✗	✗	✗	$n=3$								$3<n<\infty$
$3C_7{:}\tilde{\mathcal{P}}^-$	✗	✗	✗		✗								
$3C_6{:}\tilde{\mathcal{P}}$	✗	✗				✗							
$3C_{11}{:}\widetilde{S(3)}$	✗				✗	✗		✗		✗			
$3C_{12}{:}S(4)$	✗	✗					✗		✗		✗	✗	
$3C_{8:n}\ (3C_{10})$	✗	✗			$n=2$								$2<n<\infty$
$3C_9$		✗			✗								
$S*S$													✓
$S*H_n$	$n=3$				$n=3$								$3<n<\infty$
$\widetilde{\mathcal{P}(3)}$													✓
$\widetilde{H_{\mathcal{T}}}$	?				?								usually

TABLE 22.2. Primitive homogeneous 2-multi-tournaments: triangle constraints.

I.D.	Non-triangle constraints
$C[S]$, $Q[S]$, $\infty[S]$	$(IC_3)^1$, $(C_3I)^1$
$S[C]$, $S[Q]$, $S[\infty]$	$(IC_3)^2$, $(C_3I)^2$
$S[S]$	$(IC_3)^1$, $(C_3I)^1$, $(IC_3)^2$, $(C_3I)^2$
Q^n	Order $(n+1)$ $\overset{2}{\longrightarrow}$-tournaments $(2 < n < \infty)$
S^n	Order $(n+1)$ $\overset{2}{\longrightarrow}$-tournaments $(2 < n < \infty)$, all forms of IC_3, C_3I
$\infty\hat{*}Q$, $\infty\hat{*}\infty$	$IC_3(212;221)$, $L_4(212;21;2)$, $C_4(2121;22)$, $C_3I(221;212)$
$\infty\hat{*}S$	$IC_3(212;221)$, $L_4(212;21;2)$, $C_4(2121;22)$, $C_3I(221;212)$; $(IC_3)^1$, $(C_3I)^1$
$n*Q$, $n*\infty$	Order $(n+1)$ $\overset{2}{\longrightarrow}$-tournaments $(2 < n < \infty)$
$n*S$	Order $(n+1)$ $\overset{2}{\longrightarrow}$-tournaments $(2 < n < \infty)$; $(IC_3)^1$, $(C_3I)^1$
$Q*S$	$IC_3(\overset{b}{\longrightarrow})$, $C_3I(\overset{b}{\longrightarrow})$, where $\overset{b}{\longrightarrow}$ is $\overset{1}{\longrightarrow} \cup \overset{2}{\longleftarrow}$
$Q*H_n$	Order n cliques in $\overset{b}{-} = \overset{1}{\longrightarrow} \cup \overset{1}{\longleftarrow}$; i.e., order n tournaments in $\overset{1}{\longrightarrow}$.
$3C_{8:n}$	$L_{n+1}(1)$ $(2 < n < \infty)$
$S*S$	IC_3, C_3I for the relations $\overset{a}{\longrightarrow} = \overset{1}{\longrightarrow} \cup \overset{2}{\longrightarrow}$ and $\overset{b}{\longrightarrow} = \overset{1}{\longrightarrow} \cup \overset{2}{\longleftarrow}$
$S*H_n$	IC_3, C_3I for $\overset{1}{\longrightarrow} \cup \overset{2}{\longrightarrow}$, order n for $\overset{1}{\longrightarrow}$ if $3 < n < \infty$
$\widetilde{\mathcal{P}(3)}$	Same as $\mathcal{P}(3)$ (order 4)
$\widetilde{H_{\mathcal{T}}}$	non-triangles of $\mathcal{T}$, if any (same as $H_{\mathcal{T}}$)

TABLE 22.3. Nondegenerate homogeneous 2-multi-tournaments: non-triangle constraints.

Appendix B

OPEN PROBLEMS AND SOME RECENT RESULTS

The discussion in Appendix A, at the end of Volume I, deals mainly with recent advances relating in various ways to the material of Volume I. Here our focus shifts to open problems (or recent advances) in the theory of homogeneous structures generally, and related subjects.

We do not address the very rich theory of *finite covers* of countably categorical theories. An excellent source for an extended discussion of a number of problems in the theory of homogeneous structures is Macpherson [2011].

We also say very little about the highly developed interaction between Fraïssé theory and the algorithmic study of constraint satisfaction problems. A very comprehensive treatment of model theoretic, algebraic, and algorithmic approaches in the study of constraint satisfaction problems, notably constraint satisfaction problems with $\aleph_0$-categorical templates, is found in Bodirsky [2021], which contains a list of 59 open problems in that area.

For a discussion of the classification of homogeneous structures in a language with finitely many linear orders, and for further remarks on the classification of metrically homogeneous graphs, and its relationship to the theory of generalized metric spaces, see Appendix A in Volume I. We also dealt there with some combinatorial problems arising from the study of automorphism groups, purely from the combinatorial side. Here we review the connections of the combinatorial work with the study of the automorphism groups, as it constitutes an essential part of the general theory as well as the primary motivation for much of the recent work on the combinatorial side.

Our discussion here is keyed to the following list of topics.
– *Model theoretic problems* (*B*.1)
B.1.1 Structural Ramsey theory, partial structures, and applications
B.1.2 The finite case
B.1.3 Model theoretic simplicity, pseudo-planes
B.1.4 Homogeneous structures in ternary languages
B.1.5 The finite model property; finite axiomatizability
B.1.6 Homogenizability
B.1.7 n-Cardinal properties
B.1.8 Partial orders of width n: model companion

- *Automorphism groups* (*B*.2)

- *Algorithmic problems* (*B*.3)

- *Miscellaneous* (B.4)

B.1. Model theoretic problems

B.1.1. Structural Ramsey theory, partial substructures, and applications. For structural Ramsey theorists, amalgamation is both a strong consequence of the Ramsey property (under mild hypotheses) and a tool for the proof of structural Ramsey theorems (Nešetřil [1979]).[6]

B.1.1.1. *Existence of Ramsey expansions.* The outstanding question in this vein is the following, from Bodirsky, Pinsker, and Tsankov [2013], Melleray, Nguyen Van Thé, and Tsankov [2016].

PROBLEM 17. *Given an amalgamation class in a finite language determined by finitely many constraints—*

17.1 *Is there another such amalgamation class in an expanded language, with the Ramsey property?*

17.2 *—and is there a model theoretic or algorithmic computation providing a canonical expansion of this type?*

A similar question makes a good deal of sense, a priori, for countably categorical theories as well. But this formulation turns out to have a very interesting refutation at that level of generality, using Hrushovski amalgamation (Evans, Hubička, and Nešetřil [2019]). This tends to suggest the more restricted formulation is also dubious, and at best elusive. But as it holds in all known cases, even if it does turn out to be false in general there is something to be accounted for.

---

[6]The usual Ramsey theorem is a uniformization property for colorings of finite ordered sets; one may ask whether a similar property is shared by other classes of structures. I heard Erdős raise this question in its most concrete instance for K_6-free graphs in a lecture in the spring of 1966, or thereabouts, though it was quite some time before I realized what he was getting at.

Taking a very different approach, the preprint Hrushovski [2020] shows that any theory admits a canonical minimal Ramsey expansion. In principle this would appear to offer a valuable perspective on the question. Whether it can actually be brought to bear concretely on these problems remains to be seen.

In a different direction, the following problem has a similar flavor.

PROBLEM 18. *Given an amalgamation class in a finite language determined by finitely many constraints, is there an expansion by a finite number of sorts with elimination of imaginaries? If so, can the expansion be found in a straightforward, or at least algorithmic, manner?*

B.1.1.2. *Combinatorial methods.* There are a number of practical techniques for establishing the Ramsey property in concrete cases (and, more recently, not so concrete cases, as well). One such is the so-called *partite method*. A sort of "user interface" to the method can be found in Hubička and Nešetřil [2019]. Namely, if one can find a suitable presentation of the structure (involving the choice of language and possibly the addition of new sorts) and establish a certain general finiteness property, then the partite method can be invoked as a "black box."

A quite special case of this is the finiteness property we discussed in the appendix to Volume I, involving partial substructures. In view of its importance as an illustration of the general theory we repeat the relevant definitions.

In the combinatorial point of view the focus is on amalgamation classes $\mathcal{A}$ of finite structures rather than their Fraïssé limits, and, in practice, more particularly the associated classes $\hat{\mathcal{A}}$ of *partial substructures* (for model theorists, "weak substructures") where one may cut down not only the domain of the structure but the relations imposed—a practice which is normal in graph theory, less normal in working with linear orders. Then what is of interest is not so much the characterization of $\mathcal{A}$ by forbidden induced substructures, but the characterization of $\hat{\mathcal{A}}$ by forbidden partial substructures.

In the metric setting, we characterize the class $\mathcal{A}$ by the triangle inequality; we characterize the class $\hat{\mathcal{A}}$ by forbidding non-metric cycles of arbitrary length. Fortunately if the set of distances involved is finite, then the lengths of the cycles are bounded a priori and there are again a finite number of constraints. (On the other hand if one specializes the metric case further, e.g. to the ultra-metric case, this is no longer true.)

To verify the stated characterization of $\hat{\mathcal{A}}$ one gives a *completion procedure*: e.g., one may extend a partial metric space of diameter at most δ to a metric space of diameter at most δ by the shortest path metric truncated to δ, and one can then read off the forbidden partial substructures as the obstructions to success of the algorithm (failure consists of an alteration of one of the values of the partial metric as given initially).

The finiteness condition imposed in Hubička and Nešetřil [2019] is considerably more flexible than the one we have just given, but the simple version stated here suffices for many applications; in particular it is applicable to most of the known metrically homogeneous graphs of generic type, as discussed in detail in Volume I. We should note that we are also restricting ourselves here to strong amalgamation classes, though this condition is weakened (by allowing for algebraic closure) in Hubička and Nešetřil [2019].

As we discussed in Volume I, early work toward the finiteness property for known metrically homogeneous graphs of generic type formed a part of the thesis Coulson [2019], under substantial numerical restrictions on the numerical parameters associated with the graph. A breakthrough in Aranda et al. [2021] gave more or less the general result. This is slightly confusing; our citation refers to the short published version, where for the sake of simplicity some numerical restrictions were still imposed, but were not essential. The full version (at 57 pages) giving the general case was written about the same time, but remains a preprint (Aranda et al. [2017]). It includes some subtle variations as far as the applications are concerned, in extreme cases. In what follows we refer to the extended preprint, and not all points mentioned are touched on in the published version.

So the core issue, modulo general theory, is to find an explicit completion procedure for partial structures, which should also be canonical (invariant under isomorphism). Only later did it appear that these completion procedures could be interpreted as "shortest path completion" (generalized path metric) in the context of generalized metric spaces with values in an exotic partially ordered commutative semigroup. Such generalized metric spaces had been studied in the linearly ordered case by Conant, beginning with Conant [2015], [2017], and in the case of lattices in Braunfeld [2016]; the spaces in the latter case are *generalized ultra-metric spaces* in a sense known previously in computer science, and occasionally used elsewhere.

The idea that the conventional metric spaces afforded by metrically homogeneous graphs are better viewed as generalized metric spaces is unexpected and immensely clarifying. A forerunner is in Konečný [2019a] in the older language (where the details are checked in all cases, without unnecessary numerical restrictions) and the new point of view is exploited throughout in Konečný [2019b]. One hopes to see a systematic account in Hubička, Konečný, and Nešetřil [2020a] as well, when available.

Viewing metrically homogeneous graphs as generalized metric spaces involves putting an exotic semigroup structure on the set of distances $[\delta]$. In retrospect addition truncated to $[\delta]$, i.e., $a \oplus b = \min(a + b, \delta)$, is an example of this. In that particular case the associated partial ordering is the usual total ordering. Typically the addition is considerably stranger and the ordering is not even total. We discussed these examples in Volume I.

For further reference see Evans et al. [2019], where the subject involves stationary independence relations and the canonical completion process (as an amalgamation procedure) is very much at the heart of the matter.

There are some indications that the theory of generalized metric spaces can explain the structure of infinite primitive homogeneous binary structures with symmetric 2-types (and I suppose there should be a variant allowing anti-symmetry, though this is a mystery). Konečný made some computer explorations but this is a subject that is not very clear. One problem is that the representation by generalized metric spaces is far from canonical (as illustrated by the discussion of the "neutral" parameter in the appendix to Volume I) and we have no idea how to make it canonical. If one were going to prove something one imagines it would involve some canonical construction.

More precisely, there are two questions.

PROBLEM 19.

19.1 *Classify the 3-constrained homogeneous structures in finite binary symmetric relational languages.*

19.2 *Classify the infinite primitive homogeneous structures in finite binary symmetric languages.*

One imagines that the answer to the first question might involve generalized metric spaces and one can conjecture that the known method for going from the 3-constrained case to the general case via a suitably general notion of Henson constraint explains the rest.

B.1.1.3. *Applications.* Ramsey's original theorem was the translation into combinatorial terms of a definability theorem for the rational order, proved with a view to its application to a decision problem. Structural Ramsey theoretic results provide a useful and versatile tool for model theorists and complexity theorists in many areas: dynamical and descriptive set theoretic properties of automorphism groups of homogeneous structures, the classification of reducts of homogeneous structures, and the complexity of constraint satisfaction problems. In addition, the completion procedures for partial structures which come into play in proving finiteness theorems for forbidden partial structures have important applications to the study of automorphism groups which do not pass through the Ramsey theory.

As we will discuss properties of automorphism groups and some algorithmic issues later, for the present we focus on the classification of reducts. Actually, this is also a problem about automorphism groups as well: we want to classify reducts up to first order definability, which amounts to classifying closed overgroups of the automorphism group within the full symmetric group, viewed as a Polish group. Furthermore, with some modifications the topic has applications to algorithmic problems, as well.

Simon Thomas has asked the following.

PROBLEM 20. *Does every homogeneous structure in a finite relational language have finitely many reducts up to interpretability?*

Many instances of this are known. The most effective approach, in general, is to find a Ramsey expansion and then solve the "harder" problem of finding all reducts of that expansion explicitly. Or, more precisely, try to identify the join irreducibles in the lattice of reducts first, and then argue that all reducts are in fact joins of these.

Here one does, in fact, work with automorphisms, and one uses a Ramsey theoretic argument to canonize their form. A survey of that approach may be found in Bodirsky and Pinsker [2011].

The details of the analysis become quite elaborate. One has, for example, 39 reducts of the random permutation (Linman and Pinsker [2015]), 42 reducts of the random ordered graph (Bodirsky, Pinsker, and Pongrácz [2015]), and 116 distinct reducts of the rational order when a constant is added naming one element (Junker and Ziegler [2008]).

We have mentioned (in Volume I) a different approach to the problem in the context of homogeneous multi-orders due to Simon [2021]. When applicable, it is considerably less onerous to carry through.

When one comes to algorithmic problems, the issue shifts to the classification of reducts up to positive existential interdefinability, as this is relevant to complexity reductions. Ramsey theoretic methods apply in that setting as well.

B.1.2. The finite case. Lachlan developed an influential theory which can be viewed either as the classification of the finite homogeneous structures in a fixed finite relational language, or the classification of the stable homogeneous structures in such a language. The finite ones arrange themselves in finitely many families associated with certain numerical invariants, and the stable ones allow some or all of these parameters to go to infinity. (For a generalization beyond the homogeneous context, also suggested by Lachlan, see Cherlin and Hrushovski [2003].)

Every finite structure is homogeneous for a finite relational language. This amounts to looking at structures from the point of view of permutation group theory, in terms of their automorphism groups and the invariant relations under these groups.

One may view the restriction to finite relational languages in Lachlan's theory as imposing two constraints: (1) a bound on the *relational complexity* of the structure (or permutation group), which is the least value r such the structure is homogeneous for a relational language with relations in at most r variables along with (2) a bound on the number of relations of that complexity.

The relational complexity is an interesting parameter on its own—an invariant of permutation groups which is absent from the classical theory, and with some not very precise relationship to other classical invariants. To arrive at tractable questions on relational complexity one tends to focus on primitive structures.

I conjectured, on the basis of some computations made in the Cayley system (a forerunner of GAP) back in 1989, that the homogeneous finite primitive structures with a binary relational language are the following.[7]

(a) p-Cycles for p prime (both oriented and symmetric);

(b) Affine planes equipped with an anisotropic quadratic form; and

(c) The set with no structure (as a permutation group, this is the symmetric group acting naturally).

This has just now been completely proved via elaborate developments involving a close look at primitive actions of almost simple permutation groups.

The work in Cherlin [2016], Wiscons [2016] reduces the problem to the almost simple case. The cases of sporadic, alternating, or Lie rank 1 socle are treated in Dalla Volta, Gill, and Spiga [2018], Gill and Spiga [2016], Gill, Hunt, and Spiga [2017]. This leaves exceptional Lie type and the various Aschbacher classes associated with classical groups, now the subject of a monograph by by Gill, Liebeck, and Spiga [2021]. There is an erratum associated with the reduction to the almost simple case, which should appear in that monograph as well.

There remain a wide variety of group theoretic and combinatorial problems associated with the subject of relational complexity, qualitative and quantitative. In particular one has questions of the following types.

1. Determine the relational complexity exactly, or within close bounds, for particular actions of interest.

2. Characterize families of primitive actions modulo a bound on relational complexity (loosely).

3. Give a good heuristic for predicting relational complexity of primitive actions. Classify those with an unusually high or low complexity modulo the heuristic.

4. Find the mathematical content of the canonical language for structures of interest.

The second question is simply the most extreme and possibly the most attractive special case of the third question.

B.1.2.1. *Calculating relational complexity.* I've been working with Josh Wiscons, off and on, on various aspects of this problem, including the development of software tools in GAP for evaluating relational complexity in small structures (typically of size less than 200). As there are built-in libraries of primitive actions, this generates quite a bit of data, which I think is revealing, but this is not the place to go into the details.

One thing we noticed is that in addition to the relational complexity the *relational complexity spectrum* can be interesting and clarifying as well. Namely,

---

[7] I believe my first public mention of the conjecture was in 2000, in Cherlin [2000].

the relational complexity is the largest r for which $(r-1)$-types do not determine r-types in a straightforward way (this should not be confused with notions like $(r-1)$-closure). The associated *spectrum* is the set of all such r. Working out the relational complexity precisely is likely to involve methods relevant to the determination of the spectrum, and the latter may explain more fully what one finds—in the most straightforward cases the spectrum is an interval, but where there are gaps there are sometimes various mechanisms in play at different points.

The precise computation of relational complexity in actions involving symmetric groups tends to lead to quite specific problems in combinatorics (other cases may lead to similar problems, but in linear algebra). An early and spectacular computation of such a computation concerns wreath products of symmetric groups acting naturally (with the product action). We may call this n^d, or perhaps $[n]^d$, with d the number of factors, n the degree of the symmetric group. This is given in Saracino [1999], [2000]. The precise result is hard to state explicitly (this has probably never been attempted in public). We can however give the problem in an equivalent form for which the statement remains technical but reasonable. Namely, one views the relational complexity r as a function of a fixed n and a variable d, increasing with d, and one asks (more or less) for the value of d which first produces a given value of r, calling this value $\delta(r, n)$. The formula for δ is tricky—it measures which of several possible constructions is most efficient in a given case, Since the formula for δ is delicate, giving an explicit formula for r—essentially, the inverse function—is troublesome. We will not even reproduce the formulas for δ here—they may be found in Saracino's papers, along with a more precise and slightly more subtle definition of the function in question.

There are many other interesting questions involving wreath products, and one hopes that Saracino's methods will some day be developed farther to answer them, exactly or qualitatively. The most obvious instance would involve the action of a wreath product of symmetric groups on $\begin{bmatrix} n \\ k \end{bmatrix}^d$, where $\begin{bmatrix} n \\ k \end{bmatrix}$ refers to k-sets in $[n]$. A considerably more accessible question in this vein would be the relational complexity (or actually, the associated spectrum) for the natural action of a wreath product of alternating groups on the product space $[n]^d$.

The relational complexity of the action of symmetric or alternating groups on k-sets is known precisely. The next natural case would be the action on partitions of fixed shape, which is challenging (and seems to require a better understanding of wreath product actions on k-sets). With Wiscons we have worked out the relational complexity for the case of shape $n \times 2$ (meaning there are n pieces, each of size 2), for the symmetric group and the alternating group—n in the former case, and $n-1$, n, or $n+1$ in the latter case. On the other hand we have not managed to work out the relational complexity of the action of $\mathrm{Sym}(2n)$ on partitions of shape $2 \times n$. We offer some data.

Shape	Degree	R.C.
2×2	3	2
2×3	10	3
2×4	35	5
2×5	126	4
2×6	462	6
2×7	1716	$\geq 5; \neq 6$

This is non-monotonic in n and it is reasonable to think that parity—or even the binary representation of n—is significant.

Passing momentarily to linear algebra, the relational complexity of the general linear group acting naturally in dimension d is usually $d + 1$ (the field of order 2 provides an exception) and this is entirely to be expected. On the other hand the relational complexity of the affine orthogonal group acting naturally in the anisotropic case is 2, as the values of a quadratic form on differences of vectors determines everything, including all linear relations. (Over finite fields there are not so many examples but this all makes sense over infinite fields, as well, allowing infinite structures and an infinite language.)

The 1-dimensional affine semi-linear group $A\Gamma L(1, q)$ acting naturally raises a delicate question as to its precise relational complexity. Given the low dimension, it presents few issues qualitatively (though the Galois action on the field complicates matters). One can show on quite general grounds that in fact the relational complexity in such cases is 3 or 4, and for many purposes we will not much care which it is. However actually settling this last point seems to involve obscure Galois theoretic complications. One should note in this connection that for affine groups, the relational complexity is more closely connected with the dimension when the group is viewed as a semilinear group rather than as a linear group (and certainly the dimension over the prime field is not very relevant).

The only values for which the relational complexity of $A\Gamma L(1, q)$ is known at present to be 3 are $q = p^d$ with p prime and $d \leq 3$, and $q = 2^d$ with, mysteriously, $d = 4, 6, 7,$ or 11. We have checked that the relational complexity is 4 for $p > 2$ and $d = 4$, and also in the remaining individual cases up to 2^{14}, 3^9, and 5^6 (using inefficient methods in the last cases, which are group theoretic rather than field theoretic, but have the merit of being readily available).

Similar problems affect precise computations of the relational complexity of higher dimensional affine semi-linear groups, but qualitatively, the relational complexity is close to the dimension.

B.1.2.2. *Estimating relational complexity; exceptional cases.* We consider our second and third points together. To estimate the relational complexity presumably should involve taking the O'Nan-Scott-Aschbacher theory as a point of departure and developing a different theory for each major case. But as far as primitive actions of almost simple groups are concerned, it seems to me that a plausible heuristic for the relational complexity is something like the following, which provides an upper bound.

2 + the width of the subgroup lattice of the stabilizer of a point.

Here we add 1 to account for the passage to a point stabilizer, and then take the width plus 1 of the associated subgroup lattice as a general bound for relational complexity of any action. What we are saying is that typically, after passing to a point stabilizer, there is no reason to expect the relational complexity to be much lower than the maximal possible value. Of course the point stabilizer itself may be much smaller than the ambient group. For all but the nicest actions, we would not expect the width of the subgroup lattice of the ambient group to be relevant. A more relevant lattice for the group as a whole is the lattice of all pointwise stabilizers, but we are looking for purely algebraic invariants. And what we are suggesting is that in many case the lattice of point stabilizers inside the stabilizer of one point will be very rich.

Given such a heuristic, one then wants to look not only at the case of bounded relational complexity, but more generally at the case of relational complexity much less than the value provided by the heuristic. To say anything more requires going into the various families of actions (and possibly modifying the lattice of subgroups considered in some geometrically natural cases). In principle there is also a "high side" if one uses a heuristic which serves as an upper bound. One does not expect it to be achieved, or even approximated, very often, so exceptional cases may be of interest. Given the strong possibility that the group and the point stabilizer have the same relational complexity, one should probably start with the case of near equality (within 1).

When the actions are geometrically natural the stabilizer subgroups are geometrically natural as well, and the estimate would be significantly reduced by taking this sort of thing into account. So, as we have mentioned, one should certainly refine this by taking the O'Nan-Scott-Aschbacher classification into account at the outset. Perhaps at this point the permutation theorists should weigh in, and take over.

In fact, this is what has transpired in the binary case. That work takes advantage at certain points of some rather extreme consequences of binarity. Other parts of the analysis amount to the estimation of relational complexity in natural families where the value tends to infinity for more or less transparent reasons. In such cases this already gives information about bounded relational complexity—it is less clear how much one can say at present about heuristics for the qualitatively correct value.

What one would hope to see in the case of bounded relational complexity is: small members of natural families (e.g., low dimensional actions), arbitrary actions with small point stabilizers, and perhaps a handful of interesting families. The first case may look like a subcase of the second, but we are measuring in one case by something like dimension, and in the other case with something more like the width of the full subgroup lattice.

In the binary case, we would view the anisotropic affine orthogonal groups and the natural actions of symmetric groups as falling in the third, interestingly exceptional, category. This is obscured by the bound on dimension in the orthogonal case. But one does not actually have to restrict oneself to finite permutation groups and finite fields, and then the family has unbounded dimension.

For the symmetric groups acting naturally, the low relational complexity really is a property of the structure. For the very similar action of the alternating group, the complexity jumps to $n - 2$—and the kinds of invariants we would like to work with do not make much distinction between the two groups.

If one wants a very specific description of primitive ternary groups, then as mentioned the 1-dimensional semi-linear groups present apparently delicate questions, but we would just place these on the low-dimensional side and look for more significant families of exceptional cases.

We mention also that in the very particular case of binary permutation groups, it is not clear that a classification necessarily requires primitivity.

On the one hand, we have the difficulty that the imprimitive binary case will not reduce in any direct way to the primitive binary case. For example, every regular action of a group is binary, and every transitive action is a quotient of a regular one.

But, on the other hand, there may still be a structural analysis incorporating primitive and regular actions as building blocks lying at opposite extremes. This subject lies outside the usual range of permutation group theory but is normal enough for model theory. Algebra aims mainly to understand the basic structures, and model theory tends to study how they can interact. I don't expect model theory to account for the structure of an arbitrary finite group, but if one adds the regular actions of finite groups to the list of "basic geometries" one can then look farther.

The base case for this would be height 2 in the sense that the quotient by a minimal invariant equivalence relation is primitive. As we have no information on the quotient, a priori, this seems like a large problem. Consider the simplest case of degree $2p$ with p prime. Then for $p > 3$ there are 11 known examples arising from primitive and regular groups by taking compositions and products; and also a further example for $p \equiv 1 \pmod 4$ arising from a sort of twisted isomorphism of between components. For primes $p \leq 13$ there are no other transitive binary structures of order $2p$ and it is not unreasonable to expect this to hold for general p, and, in general, for some structural description in the general case of transitive binary structures. One has to take into account at least the possibilities of forming compositions, possibly twisted by automorphisms up to a permutation of the language, and products.

B.1.2.3. *Invariant relations.* The subject of relational complexity concerns a general form of invariant theory. For example, for the natural action of the general linear group (or its affine version if we insist on primitivity), the relations relevant to the determination of the relational complexity are the linear relations, while for the orthogonal group the linear relations may be dropped when the form is anisotropic. In the case of the projective line the relational complexity is 4, and this is witnessed by the invariant relations given by the cross ratio.

One can also work out the relational complexity and corresponding invariant relations for interesting actions of sporadic groups, and the question remains as to whether they have mathematically meaningful content, notably in highly transitive cases.

B.1.3. Model theoretic simplicity, pseudoplanes.

PROBLEM 21 (Koponen [2016]). *Is every simple homogeneous structure for a finite relational language supersimple of finite rank?*

THEOREM B.1.1 (Koponen [2016]). *A simple homogeneous structure in a finite binary language is supersimple of finite rank.*

THEOREM B.1.2.

1. (Koponen [2017]) *A primitive homogeneous binary structure with a simple theory is a random structure.*
2. (Palacín [2017]) *Any 2-transitive finitely homogeneous structure with a supersimple theory satisfying a generalized amalgamation property is a random structure.*

An obstacle to stating a more general conjecture as a problem is the difficulty in accounting for interpretations and homogenizable reducts, particularly the latter.

PROBLEM 22. *Is there a Lachlan pseudo-plane which is homogeneous for a finite relational language?*

THEOREM B.1.3 (Thomas [1998]). *There is no Lachlan pseudo-plane which is homogeneous for a finite binary relational language.*

B.1.4. Ternary languages. Part I was sparked by Nguyen Van Thé's hope that some interesting examples relating to Ramsey theory might turn up; given the difficulties of systematic classification, this was indeed a natural place to look, though it does smack of looking for one's keys under the lamp rather than where they were lost.

If possible, one would like to search systematically for homogeneous structures in a ternary language in which no binary relations are definable.

We indicate two directions which seem natural.

B.1.4.1. *3-Hypergraphs and 3-hypertournaments.* We may fix a ternary predicate $T(x, y, z)$ and require irreflexivity: $T(x, y, z)$ implies x, y, z are distinct. It is natural to impose one of the following symmetry conditions:

(S) $T\mathbf{x} \iff T\mathbf{x}^\sigma$ ($\sigma \in \mathrm{Sym}(3)$)—symmetry: the 3-hypergraph condition;
(A) $T\mathbf{x} \iff \neg T\mathbf{x}^\sigma$ ($\sigma \notin \mathrm{Alt}(3)$)—anti-symmetry: we call this the 3-*hypertournament* condition

Some steps toward consideration of homogeneous 3-hypergraphs are found in Akhtar and Lachlan [1995]. Bearing in mind that the classification of homogeneous graphs is more elaborate than the classification of homogeneous tournaments, one might prefer to start on the anti-symmetric side.

PROBLEM 23. *Classify the homogeneous 3-hypertournaments (or, analogously, t-hypertournaments, with a similar definition of an anti-symmetric t-place relation).*

I looked at the finite case and found the following. The finite homogeneous t-hypertournaments, for $t \geq 2$, are just the *trivial* ones (on fewer than t vertices), the canonical t-hypertournament on $(t + 1)$ vertices with automorphism group $\mathrm{Alt}(t + 1)$ (the 3-cycle, for $t = 2$), and a 3-hypertournament on 8 vertices with automorphism group $A\Gamma L(1, 8)$. This relies on Kantor [1972].

I also looked at the infinite 4-constrained homogeneous 3-hypertournaments. There are only three 3-hypertournaments of order 4, and four infinite 4-constrained homogeneous 3-hypertournaments. Three of these come quickly to mind: the generic circular order, the generic 3-hypertournament, and the generic "even" 3-hypertournament, meaning that on any ordered 4-tuple an even number of increasing triples lie in the relation. The fourth one is defined by a forbidden 4-type and an amalgamation procedure, neither of which immediately suggests an interpretation of the object (though presumably there is one).

These examples have in fact served as fodder for a Ramsey theoretic investigation in Cherlin et al. [2021], where they are described more fully. My contribution was pointing out that they exist; one of them appears to be quite interesting.

PROBLEM 24. *Are there infinitely many homogeneous t-hypertournaments for some t (notably, for t = 3)?*
For $t \geq 3$, are there any that are not $(t + 1)$-constrained?

No doubt the problem of classifying the $(t + 1)$-constrained homogeneous t-hypertournaments is attractive, useful, and tractable as well.

If one wants to tackle the characterization of the generic homogeneous 3-hypertournament under the assumption that all 3-hypertournaments of order 4 embed, or some stronger hypothesis as appropriate, then one would begin by thinking about the Lachlan Ramsey-theoretic method. It seems inapplicable here. One makes use there of the characteristic feature of binary structures, namely that in order to determining the type of some elements over an arbitrary set A of parameters, when the 1-types over A are already known, is a matter of settling the types of the new elements among themselves, and the base set plays no further role. One can certainly contemplate making some adjustment of the method to avoid this point, but at the essential step where one uses Ramsey's theorem to control the structure of an amalgam, this seems implausible.

B.1.4.2. *Homogeneous families of linear orders.* A structure (A, T) with T ternary and irreflexive may be viewed as a family $(A \setminus \{a\}, T_a)$ with the parameter a varying over A, where $T_a(x, y) \iff T(a, x, y)$. In particular a *cyclic order* is a 3-hypertournament (A, T) for which the associated relations T_a are linear orders.

Setting aside the trivial case $(|A| < 3)$, a homogeneous cyclic order is universal (dense).

Now instead of 3-hypertournaments let us consider irreflexive ternary structures (A, T) for which the corresponding structures $(A \setminus \{a\}, T_a)$ are linear orders. We call such a structure a *family of linear orders* (FLO).[8]

PROBLEM 25. *Classify the homogeneous FLOs.*

We have the generic cyclic order, and at the opposite extreme, the generic FLO. We will see that there are uncountably many homogeneous FLOs, which does not necessarily preclude their classification.

There are two isomorphism types of FLO of order 3, corresponding to the isomorphism types of tournaments of order 3: we call them *linear* or *cyclic* correspondingly. The cyclic orders are the ones omitting the linear 3-type; the ones omitting the cyclic 3-type also form an amalgamation class, so there is a *generic anti-cyclic FLO*.

Now we describe uncountably many homogeneous FLOs.

DEFINITION B.1.4. A ternary relation R is *cyclically invariant* if it is invariant under cyclic permutations of the variables (hence, on each triple, either fully symmetric or anti-symmetric).

The *cyclically invariant part* T_{cyc} of a ternary relation T is the largest cyclically invariant relation contained in T.

Let $\mathcal{F}$ be a family of finite ternary structures (A, R) with R irreflexive and cyclically invariant. An FLO $\Gamma = (X, T)$ will be said to be $\mathcal{F}$-*free* if (X, T_{cyc}) does not contain any structure in $\mathcal{F}$ as a *weak* substructure.

Let $\mathcal{F}$ be a set of irreflexive cyclically invariant ternary structures (A, R) which are irreducible in the sense that every pair of vertices belongs to a triple in R, Then we claim that the class of finite $\mathcal{F}$-free FLOs is an amalgamation class. We also claim that there are uncountably many amalgamation classes of this form.

---

[8]In connection with Ramsey theory it is natural to consider ternary structures (A, T) which are generic subject to the restriction that the structures $(A \setminus \{a\}, T_a)$ lie in a fixed strong amalgamation class. One then expects FLOs to play a role in the Ramsey theory; but one must also have a global linear order defined without parameters.

To check amalgamation, we consider a 2-point amalgamation problem $A_0 \cup \{a_1, a_2\}$ where the structure of $A_1 = A_0 \cup \{a_1\}$ and $A_2 = A_0 \cup \{a_2\}$ is given. For $a \in A_0$ we amalgamate the orders $<_a$ defined on $A_1 \setminus \{a\}$ and $A_2 \setminus \{a\}$, arbitrarily. We also set

$$A_0 <_{a_1} a_2; \qquad\qquad A_0 <_{a_2} a_1.$$

This construction ensures that no cyclically invariant edge containing a_1 and a_2 appears, so if a structure in $\mathcal{F}$ weakly embeds in the cyclic part of the amalgam, it must embed in a factor. Thus this class has amalgamation.

Finally, we require an infinite antichain (with respect to weak embedding) of finite cyclically invariant irreflexive and irreducible ternary structures. For any tournament T let $T^+ = T \cup \{a\}$ be given the cyclically invariant ternary structure generated by triples (a, x, y) with (x, y) an arc of T.

Then T^+ is irreflexive, cyclically invariant, and irreducible. Since every triple in T^+ contains a, if we have a weak embedding of T_1^+ into T_2^+ with $|T_1| \geq 3$ then this induces an embedding of T_1 into T_2. As there is an infinite antichain of tournaments we have the desired infinite antichain of ternary structures.

The following is a natural first question concerning the range of this construction.

PROBLEM 26. *If a homogeneous FLO is anti-cyclic (i.e., has trivial cyclic part) is it the generic anti-cyclic FLO?*

B.1.5. The finite model property and finite axiomatizability. A structure has the *finite model property* if it is a model of the theory of all finite structures. Such structures are called *pseudofinite*.

The following interesting problem is long-standing, and representative of a wide range of problems.

PROBLEM 27. *Is the generic triangle free graph pseudo-finite?*

The broader problem is the following.

PROBLEM 28. *Give an algorithm to determine whether a homogeneous structure determined by a specified finite set of minimal constraints is pseudo-finite.*

In Cherlin [2011b] we pointed out some very concrete instances of the question of pseudo-finiteness of the generic triangle free graph which had resisted analysis. One of these questions was quickly answered by Even-Zohar and Linial [2015] using a construction previously considered in connection with Ramsey theory. The other questions of this type, very similar in flavor, remain untouched.

At the opposite extreme from pseudo-finiteness is finite axiomatizability. More generally one may consider quasi-finite axiomatizability, which allows an axiom of infinity (or perhaps several) in addition to finitely many first order axioms.

PROBLEM 29 (Macpherson [1991]). *Does every finitely axiomatizable count-ably categorical theory have the strict order property?*

The problem is also of interest when specialized to homogeneous structures for finite relational languages.

THEOREM B.1.5 (Macpherson [1991]). *Every quasi-finitely axiomatized complete theory with trivial algebraic closure either has the strict order property or is a definable expansion of the theory of equality.*

A natural strengthening of the finite model property is the finite *submodel* property. The theory of the successor function (or successor relation) on $\mathbb{Z}$ illustrates the distinction.

PROBLEM 30 (Macpherson [2011, 3.2.2]). *Is there a pseudo-finite, countably categorical theory which does not have the finite submodel property?*

B.1.6. Homogenizability. A relational structure Γ in a finite language is *homogenizable* if it has a definable expansion by finitely many relations to a homogeneous structure.[9]

Covington gives a sufficient condition for homogenizability in Covington [1990]. The subject crops up in various places but might benefit from more systematic investigation.

In the study of reducts of homogeneous structures, in addition to structural Ramsey theory it seems that their classification by Ramsey theoretic methods depends on the reducts being homogenizable. However, as was pointed out by Macpherson and mentioned by Thomas [1991], not all such reducts are themselves homogenizable (in a finite relational language). A natural example of this is provided by a countable existentially complete bowtie-free graph, which becomes homogenizable when expanded by a suitable linear order (a remark of Rehana Patel and Jan Hubička, and one which is fortunate from the point of view of Ramsey theory). In this particular case the source of the phenomenon appears to lie in the difference between algebraic closure and definable closure, in the absence of a linear order.

PROBLEM 31. *Let C be a finite set of connected structures and let Γ be an existentially complete C-free structure. When is Γ homogenizable?*

The prior question of course is when is the structure $\aleph_0$-categorical. This question is intimately connected with questions on the existence of countable universal graphs. We know at this point that in most cases the existentially complete structures in such a class are not in fact $\aleph_0$-categorical, and hence not homogenizable, but the range of cases in which these structures are $\aleph_0$-categorical is broad enough to make the homogenization problem interesting. There are probably very few cases indeed in which such graphs are homogenizable. The first non-trivial case in some sense, after the case of forbidden

---

[9]A mild abuse of language, perhaps.

paths, is the case of the forbidden bow-tie, where already the structure is not homogenizable.

There is an extensive literature on the topic. We mention the survey Komjáth and Pach [1991], the general theory relating universal graphs and algebraic closure given in Cherlin, Shelah, and Shi [1999], and the case studies in Cherlin and Tallgren [2007]. There is more to the theory and perhaps one can arrive at a quick answer to Problem 31 by putting it into that context.

One can state the problem equally well for other classes of structures of combinatorial interest, or more broadly. Due to the connection with the study of universal graphs the focus has been on the graph theoretic setting. The presence of an ordering would certainly change the flavor of the problem, and bring it closer to the problem of $\aleph_0$-categoricity as such. One should certainly look into these questions for directed graphs.

The questions about universality remain interesting for uncountable structures, but take on a different character. Shelah and his collaborators have had a good deal to say about that.

B.1.7. n-Cardinal properties. The problem is to determine relations between the sizes of definable subsets in arbitrary structures elementarily equivalent to a given homogeneous structure Γ.

Relatively little is known. In Cherlin and Thomas [2002] the case of the random graph was treated, and some glib and inaccurate remarks made about other homogeneous graphs (I thank Nate Ackerman for catching this).

The accurate part has been considerably generalized.

THEOREM B.1.6 (Ackerman [2012]). *Let T be the theory of a homogeneous structure Γ for a finite relational language whose associated amalgamation class has strong amalgamation. Let $p_1, \ldots, p_n$ be the complete quantifier-free types in one variable. Then for any infinite λ, and any sequence of cardinals $(\lambda_1, \ldots, \lambda_n)$ with*

$$\lambda \leq \lambda_i \leq 2^\lambda$$

there is a model of T in which p_i defines a set of cardinality λ_i.

Note that the condition of strong amalgamation guarantees that the sets defined by the p_i are infinite.

Now it is evident that there are also cases in which cardinalities are not related at all. But other questions arise, as illustrated by the following.

PROBLEM 32. *Let Γ be the generic triangle free graph and let $\Gamma_{a,b}$ be Γ with two vertices a, b named, where (a, b) is a non-edge. What are the possible sequences of cardinalities arising as above for models of the theory of $\Gamma_{a,b}$?*

As the sets A, B defined by $x - a, x \not\!\!- b$ and by $x - b, x \not\!\!- a$ respectively are not connected by edges to the set defined by $x - a, b$, it is to be expected that a gap of two exponentials can arise, something which would certainly not occur in the setting of the random graph.

There may be a completely satisfactory analysis of the general case as well, assuming strong amalgamation, but this is all that we know at present.

B.1.8. Partial orders of width n: model companion. The following more isolated problem is not closely related to our topic, as far as we know, but the solution may involve similar ideas. We state it mainly because it is simple, natural, and has been open for a considerable time, in spite of evoking some interest.

PROBLEM 33 (Bonato and Delić [1997/98]). *Does the theory of partial orders of width at most n have a model companion? If so, is it homogenizable?*

Bonato and Delić prove that the model companion exists for $n = 2$. The rest lies in shadow.

B.2. Automorphism groups of homogeneous structures

For the automorphism group as a permutation group, one natural line concerns the study of the twisted automorphism group as a finite extension. We have said what we have to say about that in the appendix to Volume I, since the case of metrically homogeneous graphs provides an interesting case study and there is as yet no general theory.

For the automorphism group as an abstract group, the emphasis is on simplicity and more generally on the normal subgroup structure. In principle, another topic under this heading would be the cofinality of the group (and some related issues); but these tend to be more appropriately treated along with descriptive set theoretic properties of the group as a Polish group.

As far as the Polish group structure is concerned, we have both the descriptive set theory (notably, generic or ample automorphisms) and the topological dynamics (notably, extreme amenability or more generally metrizability of the universal minimal flow).

All of these matters are very rich and well documented subjects with broad connections and work actively ongoing. As mentioned in the appendix to Volume I, many problems in this area were solved in the case of (most) metrically homogeneous graphs mainly by passing through a canonical completion process inspired by the theory of generalized metric spaces. But in Volume I we had put aside the precise relationship of the combinatorial work to its intended applications, which is part of the general theory. So we will come back to that subject here.

See Macpherson [2011, §4] for a comprehensive survey of the topic, which has provided much of the motivation for recent work in the area, since the appearance of Kechris, Pestov, and Todorcevic [2005].

An up-to-date and comprehensive survey of the combinatorial work, insofar as it bears on the case of metrically homogeneous graphs and a broad range of similar cases, is found in the introduction to Hubička's habilitation (Hubička [2019]).

B.2.1. Simplicity, or normal subgroup structure. This topic can be approached via a direct study of conjugacy classes in the automorphism group (which has its own interest, notably in connection with the theory of Borel complexity). Another line of attack is developed in Tent and Ziegler [2013b], [2013a]. This approach is motivated by stability theoretic methods used by Lascar which were transferred to *free amalgamation classes* in Macpherson and Tent [2011] and then given a more abstract and flexible formulation based on the existence of *canonical amalgamation*; one route to this is to treat it as a special case of *canonical completion* of partial structures.

In the presence of anti-symmetric relations, notably in the simple case of linear orders, this notion is not available in its current sense.

When the notion is available, one expects to find canonical amalgamation used in the proof of the amalgamation property. But in the case of metrically homogeneous graphs the existence of a "canonical amalgamation" was considerably less clear than the amalgamation property itself. It was supplied in an algorithmic form in Aranda et al. [2017].

A general result given in Evans et al. [2019], from the point of view of generalized metric spaces, covers the known primitive metrically homogeneous graphs of generic type. The imprimitive case consists of bipartite and antipodal graphs. In the bipartite case one has the normal subgroup fixing the partition; that is, one should name the two parts. In the antipodal case we no longer have strong amalgamation and we may not have canonical completion either. These cases merit further consideration but do not seem to fall neatly under the general theory.

PROBLEM 34. *Is the automorphism group of a known metrically homogeneous bipartite graph of generic type, with the parts named, a simple group?*

Is the quotient of the automorphism group of a known metrically homogeneous graph of antipodal type by its center a simple group?

The usual notion of stationary independence relation has a strong symmetry axiom which will fail in many natural cases. A non-symmetric version with similar applications is considered in Li [2019].

B.2.2. Dynamical properties: extreme amenability, amenability, unique ergodicity, EPPA, ample generics and consequences. Interest in the study of automorphism groups of homogeneous structures as topological groups has been invigorated by the connections between topological dynamics, model theory, and combinatorics brought out in Kechris, Pestov, and Todorcevic [2005] and has sparked a good deal of subsequent work, notably on the combinatorial side. A useful source of additional key results and notions is Kechris and Rosendal [2007].

Here we review some key notions and connections with combinatorics that come into play generally. In particular this provides motivation for the combinatorial investigations in the case of metrically homogeneous graphs described in the appendix to Volume I (as does the material of the previous section on simplicity of the automorphism group, via a different line of thought). It is a very rich and actively developing topic, so we enter into some detail.

We being by recalling the combinatorial notions, then move on to some applications.

B.2.2.1. *Combinatorial properties.* Here, via the Fraïssé theory, the focus is on amalgamation classes $\mathcal{A}$ of finite combinatorial structures, putting us squarely in the domain of finite combinatorics. The key notions are the *Ramsey property* and the *extension property for partial automorphisms*.

DEFINITION B.2.1. Let $\mathcal{A}$ be an amalgamation class of finite structures.

1. $\mathcal{A}$ has the *Ramsey property* if for A, B in $\mathcal{A}$ there is $C \in \mathcal{A}$ with satisfying the condition (in "Hungarian" notation)

$$C \to (B)_r^A.$$

That is, a coloring of copies of A in C results in a copy of B in C on which the coloring is uniform (monochromatic).

2. $\mathcal{A}$ has the *extension property for partial automorphisms* (EPPA) if for ever $A \in \mathcal{A}$ there is a "symmetrized" $B \in \mathcal{A}$ containing A, so that every isomorphism between parts of A is induced by an automorphism of B.

We apply the same terminology indiscriminately to the Fraïssé limit of $\mathcal{A}$, for the sake of brevity.

It should be noted that the Ramsey property requires a definable linear ordering on the Fraïssé limit, for reasons which are clearest via relations with topological dynamics, discussed below. In particular in the context of primitive metrically homogeneous graphs of generic type one will work with the generic expansion by a linear order (though there are similar conditions applicable to the automorphism group of the graph itself). In any case the combinatorial analysis required to prove the desired property does not entail such an expansion.

For each of these properties, the combinatorial challenge of establishing that it holds in particular cases of interest has its own extensive literature. As far as structural Ramsey theory is concerned, one of the leading methods of proof is the so-called *partite method* of Nešetřil and Rödl. The study of EPPA (sometimes called the *Hrushovski property*) was initiated by Hrushovski. Here as well there are a number of approaches in use.

Systematic approaches to both subjects—surprisingly similar in terms of the key finiteness conditions imposed—are found in Hubička and Nešetřil [2019], Hubička, Konečný, and Nešetřil [2020b], couched in terms of passage from a class with the desired property to a suitably nice subclass. Both papers serve also as surveys and mention some additional striking open problems.

As we mentioned in the appendix to Volume I, a convenient route to a very strong and simple version of the appropriate finiteness condition on forbidden weak substructures is via a suitable completion algorithm for partial structures. We recall the terminology.

A *partial substructure* (or a *weak substructure*, for model theorists) is obtained by taking a subset of the universe of a relational structure along with a subset of each relation. For graphs this is the graph theoretic notion of substructure (the model theoretic notion of substructure translates to *induced subgraph* in this context). The Fraïssé theory deals with a class $\mathcal{A}$ closed under taking induced substructures; the combinatorics goes into the associated class $\hat{\mathcal{A}}$ obtained by taking partial substructures. In particular we may impose the quite strong (and often satisfied) requirement that there be finitely many minimal forbidden partial substructures, and this gives a structural Ramsey theorem.

One route to such a finiteness condition, discussed in some detail in a concrete case in the appendix to Volume I, is the analysis of a completion algorithm for partial substructures (leading us back eventually to generalized metric spaces, in that particular case). More precisely, the minimal forbidden partial substructures appear as obstacles to the completion process. If the completion process is suitably canonical, one arrives at the finiteness condition required for EPPA.

In the case of primitive metrically homogeneous graphs of generic type (and a number of imprimitive ones), the preprint Aranda et al. [2017] gives the completion procedure and its application to structural Ramsey theory and EPPA. Subsequently the connection to generalized metric spaces appeared, and a treatment of EPPA for generalized metric spaces is found in Conant [2019].

Once the combinatorial side is under control, primarily in the form of structural Ramsey theory or EPPA, one can pass to the more concrete study of the automorphism group of the Fraïssé limit as a Polish group.

B.2.2.2. *Automorphism groups as Polish groups.* The focus here is on the dynamical properties and the descriptive set theoretic properties of the automorphism group of a countable homogeneous structure, with the Polish topology associated with its action as a permutation group on a countable set.

On the *dynamical side* one considers *flows*, i.e continuous actions of the group in question on compact topological spaces, more particularly *minimal flows*, with no invariant subflows, and the *universal minimal flow* (a minimal flow covering all others), whose uniqueness can be established. This has a tendency to be quite large in general (like the Stone-Čech compactification). However it may reduce to a point. This is a fixed point property; every flow has a fixed point. Groups with this fixed-point property are called *extremely amenable*, and tend to be considered exotic.

A key result from Kechris, Pestov, and Todorcevic [2005] establishes the equivalence of extreme amenability and the Ramsey property for the associated amalgamation class. One of the relevant flows is the action of the automorphism group on the compact space of all possible linear orderings of the structure; a fixed point is a definable linear ordering, accounting for our earlier remark on this point.

Thus $\mathrm{Aut}(\mathbb{Q}, \leq)$ is extremely amenable by the classical Ramsey theorem. On the other hand $\mathrm{Aut}(\mathbb{Q})$ (or $\mathrm{Sym}(\omega)$, if one prefers), lacking a linear order, turns out to have the space of all linear orders as its universal minimal flow. In general when one has a suitable expansion with a structural Ramsey theorem the universal minimal flow is the action on all possible expansions of the same type. The characteristic feature of this situation is *metrizability* of the universal minimal flow, a topic taken up with satisfying results in Melleray, Nguyen Van Thé, and Tsankov [2016], Ben Yaacov, Melleray, and Tsankov [2017].

On the *descriptive set theoretic side*, following the line laid down in Kechris and Rosendal [2007], one considers mainly *ample generics* (or, as this concerns an automorphism group, *ample generic automorphisms* from the point of view of the structure). This strengthens the requirement of a generic automorphism (a comeager conjugacy class in $\mathrm{Aut}(\Gamma)$) to a comeager $\mathrm{Aut}(\Gamma)$-orbit on n-tuples of automorphisms. Nothing actually depends on the group being an automorphism group, and other cases are of interest, but our interest here is of course the case of automorphism groups.

This is more or less (i.e., usually) a consequence of EPPA; the question is whether we can move from the category of structures to the category of structures equipped with a sequence of automorphisms. In particular if the EPPA is derived from a canonical completion algorithm then one gets ample generics. From this a variety of useful properties follow in various contexts—at the level of automorphism groups of $\aleph_0$-categorical structures, and sometimes more broadly.

Consequences include automatic continuity, the small index property, and uncountable cofinality (the last not actually involving the topology, a priori). There is a rich prior history, also covered in the comprehensive paper of Kechris and Rosendal, and for an explanation of these topics, and some others as well, we refer the reader to that source.

To recapitulate: in Volume I, we saw that for the known primitive metrically homogeneous graphs of generic type there is a theory of canonical completion of partial structures which leads to finiteness properties and then to the Ramsey property and EPPA for appropriate classes of finite structures, at which point the general theory takes over to give a wide variety of applications; and much the same applies to stationary independence relations, though it took a little longer to derive the corresponding applications.

The state of affairs for the known metrically homogeneous graphs as of Aranda et al. [2017] was as follows. First if Γ is one of Macpherson's tree-like graphs $T_{m,n}$, then there is no suitable (pre-compact) Ramsey expansion, EPPA fails, and there is no stationary independence relation in the strict sense. Leaving that case aside brings us to the generic case (or something similar, in diameter at most 2). In this case the results are as follows (for some reason the restriction "3-constrained" crept into the statement of Theorem 1.1; see however the comment on page 8 regarding Henson constraints)

Theorem 1.1: Ramsey expansions: They exist and are natural. In the primitive case, a generic linear order suffices; in the bipartite case one names the parts; in the antipodal case one adds a generic linear order convex with respect to the antipodality relation—but in the bipartite antipodal case with odd diameter one should take the order induced on antipodal pairs to correspond to an ordering on the parts.

Theorem 1.2: EPPA: This holds except possibly in some antipodal cases, where it holds after a further expansion of the language splitting antipodal pairs by an equivalence relation with two classes.

Theorem 1.4: Stationary independence: In the primitive case, a stationary independence relation exists. In the imprimitive case there are subtleties—we give further details below.

The first difficulties connected with stationary independence concern canonical amalgamation over an empty base (direct sum). In the bipartite case clearly there is no canonical way to correlate the two sides in a direct sum. In the antipodal and not bipartite case, the question is how to form the direct sum of two pairs of antipodal points. If (and only if!) δ is even one may use the distance $\delta/2$ as the (symmetric) "default." In this case, use of the corresponding completion procedure produces a stationary independence relation.

On the other hand, this obstacle aside, there tends to be a (local) stationary independence relation. The exceptions come in two types of antipodal graphs: bipartite of even diameter, or non-bipartite of odd diameter. Both of these require some symmetry-breaking even over a non-empty base.

The missing cases of EPPA are resolved, favorably, in Konečný [2020], following on Evans et al. [2020].

In the appendix to Volume I we also discussed why we think it likely that all of the metrically homogeneous graphs are in fact known, so that these results would settle a number of the central problems in the area.

In connection with EPPA the following notorious problem also deserves particular mention here. There is quite a general theory but some elementary cases escape, for now.

PROBLEM 35. *Does the class of tournaments have EPPA?*

B.2.3. Descriptive set theory and Polish groups. From a descriptive set theoretic point of view the following problem is also very natural.

Problem 36. *What is the Borel complexity of the conjugacy relation in the automorphism group of a known metrically homogeneous graph?*

One may also ask whether the automorphism group determines the structure in the spirit of Mati Rubin's theory of reconstruction (Rubin [1994]).

Of course, we highlight the case of metrically homogeneous graphs, where we seem, at this point, to have all the tools one would want for a good understanding of the automorphism groups.

B.2.4. Regular actions. Now we look in a different direction.

Cameron has raised the question as to which homogeneous structures Γ can be "Cayley objects" for a given group G; that is, G should act regularly on Γ.

Theorem B.2.2 (Henson [1971]). *The random graph and the generic triangle free graph admit regular $\mathbb{Z}$-actions. The Henson graphs Γ_n for $4 \leq n < \infty$ do not.*

Many groups can act regularly on the random graph. Cameron and Johnson give a sufficient condition for a representation of a group G as a group of automorphisms acting regularly on the random graph in terms of *non-principal square root sets*

$$\sqrt{g} = \{a \mid a^2 = g\},$$

where we require $g \neq 1$. Namely:

> The group G is not covered by finitely many translates of non-principal square root sets. $\hspace{2em}(P')$

Theorem B.2.3 (Cameron and Johnson [1987]). *A countable group satisfying condition P' has a regular action on the random graph.*

They made the following remark.

> It is possible to derive a necessary and sufficient condition on the countable group G for G to be embeddable as a regular subgroup of Aut (F). However, the condition is complicated to state, and we have no example to show that it really is more general than (P').

Problem 37. *Is there a countably infinite group, not satisfying the condition P', and acting regularly on the random graph?*

The purely permutation group theoretic question they asked was whether every countably infinite group has some representation as a regular subgroup of a primitive but not doubly transitive permutation group (in other words, acting regularly on a primitive structure with at least two non-trivial 2-types). A concrete instance of the problem is the following.

PROBLEM 38 (Cameron and Johnson [1987]). *Let G be the semidirect product* $\mathbb{Z} \rtimes \mathbb{Z}_4$ *with a generator of* $\mathbb{Z}_4$ *acting by inversion. Does G have a representation as a regular subgroup of a primitive but not doubly transitive group?*

In the present state of knowledge, one may even ask whether this particular group has a regular action on the random graph—one would expect this, at least, to be tractable.

This touches the surface of what could be a very broad subject. For example:

PROBLEM 39 (Cameron, Vershik; Cameron [2000]). *Which elementary abelian groups act regularly on the Urysohn graph (integer Urysohn space)?*

THEOREM B.2.4 (Cameron and Vershik [2006]).
An elementary abelian 2-group can act regularly on integer Urysohn space. An elementary abelian 3-group cannot.

B.3. Algorithmic problems

We have alluded to the use of model theoretic methods in the study of algorithmic problems (specifically, constraint satisfaction problems) above, and we return to this briefly in the present section.

But here we focus mainly on the reverse direction—the algorithmic side of combinatorial model theory (this is not *entirely* the reverse direction—the use of model theoretic methods in the theory of constraint satisfaction problems also leads to the consideration of specific model theoretic decision problems).

B.3.1. Constraint satisfaction problems. Ramsey theoretic methods are also relevant to the problem of determining whether or not a constraint satisfaction problem is NP-hard, allowing here for the case of certain infinite templates (Bodirsky [2015]). Here the important notion of definability is *primitive positive existential definability*, and in the presence of the Ramsey property one can show that the problem of determining whether a given relation is definable in this sense from finitely many others in a quantifier-free definable reduct of a given structure becomes decidable (Bodirsky, Pinsker, and Tsankov [2013]).

Fortunately, there is now a comprehensive treatment of the interaction of complexity theory, universal algebra, Ramsey theory, and the model theory of $\aleph_0$-categorical theories, in particular, the Fraïssé theory and the general theory of *core structures*, in the subject of constraint satisfaction problems (Bodirsky [2021]), to which we refer both for the theory and for a detailed discussion of a wide variety of open problems. The decision problems in question depend on a fixed "template" structure, often taken classically to be finite. The question is then whether a given finite structure in a specified language maps homomorphically into the template. For example, with template K_k this is k-colorability, since the edge relation in K_k is irreflexive (in fact, since the edge relation is the complement of equality, the effect of the choice of language is to introduce a positive symbol for a negative relation).

The *dichotomy conjecture* of Feder and Vardi states that every constraint satisfaction problem with finite template is either in **P** or is **NP**-complete. Positive solutions have been announced by Bulatov and by Zhuk.

As explained in Bodirsky [2021], a generalization of this conjecture to $\aleph_0$-categorical templates would be excessive. A central conjecture for the case of infinite templates is the following.

CONJECTURE 2 (Bodirsky, Pinsker; Infinite-domain dichotomy). *For every reduct of a finitely bounded homogeneous structure, the corresponding constraint satisfaction problem is either in P or is NP-complete.*

Here the notion of *finitely bounded* (or *finitely constrained*) refers to the condition that the associated amalgamation class of finite structures should be defined by finitely many forbidden substructures (in the model theoretic sense: i.e., forbidden embeddings). This condition is a very natural one to impose in an algorithmic context.

As may be seen in Bodirsky [2021], a very large body of theory goes over from the case of finite templates to the case of reducts of finitely bounded homogeneous structures, or even, quite often, to the case of $\aleph_0$-structures. Among other things, once the template is infinite then the topology on the automorphism group plays a role; and as the universal algebraic approach involves not only automorphisms but endomorphisms and even polymorphisms, these too must be viewed topologically.

And we refer the reader to Bodirsky's monograph for a substantive discussion of the subject.

B.3.2. Decision problems for finitely constrained classes. This subject seems to us seriously underdeveloped, though in certain particular contexts—notably, the theory of permutation patterns—a substantial amount of work has been done (cf. Ruškuc [2005]). It would be interesting to develop this material systematically for combinatorial structures of general type.

The setting for this, as is standard in the theory of permutation pattern classes, and already mentioned in connection with constraint satisfaction problems—is a class of finite structures characterized by finitely many minimal forbidden structures. Model theory often focuses on amalgamation classes (or Hrushovski variants) but we can study the finite models of a universal sentence as a combinatorial setting given by finitely many forbidden substructures, and look at properties of the existentially complete models, notably when their theory is $\aleph_0$-categorical.

The theory is simplest when the class of interest is closed under passing to partial substructures, that is the constraints may be taken to be forbidden partial substructures (e.g., graphs and subgraphs rather than graphs and induced subgraphs).

Permutations are best viewed as sets equipped with two linear orders; since the class of linear orders is not closed under partial substructure, the class of permutations is not closed under partial substructure either.

In what follows one is free to interpret the problems in either sense, though possibly with different outcomes in the two cases (dealing with either hereditary or monotone classes, in the combinatorial language).

A fundamental question at the outset, and a sufficiently difficult one, is the problem of determining when such a class has the joint embedding property.

PROBLEM 40 (Joint embedding decision problem). *Given a finite set of forbidden substructures. Determine whether the associated class of finite structures has the joint embedding property.*

This is open, and very interesting, for the case of permutation patterns, cf. Ruškuc [2005, p. 6].

For the case of finitely constrained hereditary graph classes (that is, with forbidden induced subgraphs) the problem is undecidable, and the same applies to the *joint homomorphism property* (Braunfeld [2019]). Furthermore, joint embedding is undecidable for finitely constrained hereditary classes of 3-multi-orders. But permutation pattern classes correspond to 2-multi-orders, so, as noted, this case remains open.

The joint embedding problem is also undecidable for universal Horn sentence in binary languages (announced in Bodirsky, Rydval, and Schrottenloher [2021]), via a connection with decision problems for context-free languages (exploited also in Bodirsky, Knäuer, and Rydval [2021] to study the amalgamation property).

If $\mathcal{A}$ is a hereditary class of finite relational structures, the usual model theoretic approach is to consider existentially complete models of the corresponding universal theory. In the case of amalgamation classes, we arrive at the Fraïssé limit in this way. In general, the desirable case is that in which the class of existentially complete models is axiomatizable; in that case, the corresponding theory is called the model companion. A good setting for algorithmic model theory is the case in which the initial class is finitely constrained, and, in addition, the model companion exists; we then have a range of decision problems corresponding to desirable properties of the model companion.

In particular, completeness of the model companion corresponds to the joint embedding property in the initial class of finite structures.

So one has, among others, the following natural decision problems, generally with several variations possible in the nature of the constraint and the level of generality (graphs, tournaments, permutations, general structures, etc.).

PROBLEM 41. *Is there an algorithm which determines, for a finitely constrained collection of finite structures with the joint embedding property, whose theory has a model companion T, whether the following properties hold?*

(A) T is countably categorical;

(B) T is small (the universality problem, see below);

(C) T is stable, simple, ...;

(D) T has elimination of imaginaries.

The universality problem we have in mind is the following: if $\mathcal{A}$ is a class of finite structures (e.g., graphs) let $\bar{\mathcal{A}}$ be the class of countable structures whose finite induced substructures lie in $\mathcal{A}$. The problem is to determine whether $\bar{\mathcal{A}}$ contains a universal structure Γ, that is, a structure which contains an isomorphic copy of each such structure as an induced substructure. This requires that $\mathcal{A}$ at least have the joint embedding property.

This problem has been studied in considerable detail in the context of graphs. One imposes finitely many constraints (forbidden subgraphs, or possibly forbidden induced subgraphs). To ensure joint embedding holds one may require the constraints to be connected. For classes $\mathcal{A}$ determined by finitely many forbidden *induced subgraphs* the decision problem for the existence of a universal model is undecidable (Cherlin [2011a]). But it may well be decidable in the context of finitely many forbidden (and connected) *subgraphs*. There is an extensive theory relating to this case, which is given in Cherlin, Shelah, and Shi [1999], and has been extensively applied.

In the case of finitely many forbidden subgraphs, the associated model companion T always exists, and the universality problem is equivalent to the existence of a countable saturated model of the theory T. Theories with a countable saturated model are called *small*. Thus we rejoin problem (B) above. The general theory is somewhat more concerned with problem (A), which is more natural from a model theoretic point of view.

Within the class of small theories, the $\aleph_0$-categorical ones are a very special subclass. But it turns out that on the one hand there is a good characterization of $\aleph_0$-categoricity of the theories T which arise in this setting in terms of model theoretic algebraic closure, and on the other hand it also appears, very fortunately, that for the theories under consideration here the gap between smallness in general and $\aleph_0$-categoricity is small. To illustrate: if one looks at classes of graphs defined by forbidding a single tree F as a subgraph, the case in which the associated model companion T is $\aleph_0$-categorical is the case in which the tree in question is a path, while the case in which the theory T is small (so a universal C-free graph exists) is that in which the forbidden tree F can be obtained from a path by adding at most one more leaf.

Permutation pattern classes are similar in some ways to classes of finite graphs determined by forbidden subgraphs, and are a very natural setting for algorithmic problems. Unlike classes of graphs, permutation pattern classes cannot be closed under taking *weak substructures,* and as a result in the case of permutation pattern classes we do not have a theory at all similar to the theory for graphs with forbidden subgraphs. In fact the problem in this setting may be more closely analogous to the wilder theory for forbidden *induced subgraphs*, but this is not known to be the case either.

It would be very good to have a theory of permutation pattern classes allowing countable universal (and, more particularly, $\aleph_0$-categorical) limit models, as such classes should be very well behaved. At present, in this direction we have, mainly, Cameron's classification of the *homogeneous* permutations in Cameron [2002/03], which is a natural first step in this direction.

In the literature on permutation pattern classes, one finds the very different question as to whether the finite permutations in the class coincide with the restrictions of one infinite permutation. This is very far from universality: for universality, we consider all *countable* permutations whose finite restrictions are in the given class, and look for one countable permutation in which they all embed. Unlike the purely finitistic version, this should be a powerful and restrictive condition.

Here are two problems—or two variations on a theme—which would arise naturally in this context.

PROBLEM 42. *If a hereditary class of finite permutations allows a countable universal limit permutation* Γ, *are the two orders on* Γ *dense?*

PROBLEM 43. *How does one calculate the algebraic closure operation in an existentially closed* $\mathcal{A}$*-permutation, where* $\mathcal{A}$ *is a hereditary class of finite permutations?*

I hesitate to tackle model theoretic decision problems in the context of forbidden induced structures as they have a tendency to be undecidable. But there is a lot of concrete work in that direction, notably on the Erdős-Hajnal conjecture in graph theory (and a little bit, by analogy, in the context of tournaments).

When problems are decidable, whether trivially or for subtle reasons, complexity theoretic questions arise—such as the complexity of the classification of 3-constrained structures, which is of concrete concern.

B.4. Old chestnuts

The first two problems below have only loose connections with our topic: the first via Ramsey's theorem, the second via Hrushovski constructions. But they are old chestnuts on the interface between model theory and combinatorial constructions.

B.4.1. Locally finite (semifinite) generalized n-gons.

PROBLEM 44. *Is there an infinite generalized n-gon with finite lines? In particular, can this occur for* $n = 4$?

Theorem B.4.1 (Cherlin [2005]). *A generalized quadrangle with at most 5 points per line is finite.*

The proof is typically model theoretic, relying on the existence of a model with an infinite sequence of indiscernible lines.

For the case of generalized hexagons see Bishnoi and De Bruyn [2016]. This seems difficult, and while one might expect some sort of analog of the model theoretic approach used for 4-gons—certainly one can start with generation by indiscernibles and various associated permutations—there is no obvious analog of even the first step in that analysis.

Strictly speaking, one can separate out two questions. In the case of 4-gons, one can define a general notion of *free generation by indiscernibles.* The bulk of the analysis in known cases shows there is no freely generated structure with the specified parameters. Some lemmas show that generation by indiscernibles implies sufficient freeness to make the argument general. So one might focus on the question of existence of generalized quadrangles with 6 points per line freely generated by indiscernibles, which is mainly a question about the group Sym(6)—not an easy one, it appears, but one that can also be approached via machine computation.

B.4.2. Stable countably categorical groups.

Problem 45. *Is there a stable countably categorical group which is not abelian by finite?*

Theorem B.4.2 (Felgner [1978], Baur, Cherlin, and Macintyre [1979]). *Countably categorical stable groups are nilpotent by finite.*

See also Baginski [2009].

B.4.3. Transitivity conditions. For the large subject of homogeneity through the lens of permutation group theory we refer to Cameron [1990] and much of Macpherson [2011]. But we take note of two specific problems on degrees of homogeneity. Here the terminology is group theoretic: t-transitivity is transitivity on ordered sets of distinct elements, and t-homogeneity is transitivity on unordered sets of distinct elements.

Problem 46 (Cameron [1990, p. 64]). *For $k \geq 5$, is there a $(k + 1)$-homogeneous and $(k - 1)$-transitive but not k-transitive permutation group?*

Theorem B.4.3 (Macpherson [1986]). *For $k > 3$, a $(k + 3)$-homogeneous, not k-transitive permutation group is a subgroup of the group D preserving a separation relation.*

Problem 47. (Peter Neumann, 1995; see Mazurov and Khukhro [2014, 11.69]) *A group G acting on a set Ω is said to be 1-{2}-transitive if it acts transitively on the set $\Omega^{1,\{2\}} = \{(\alpha, \{\beta, \gamma\}) \mid \alpha, \beta, \gamma \text{ distinct}\}$. Thus G is 1-{2}-transitive if and only if it is transitive and a stabilizer G_α is 2-homogeneous on $\Omega \setminus \{\alpha\}$. The problem is to classify all (infinite) permutation groups that are 1-{2}-transitive but not 3-transitive.*

REFERENCES FOR VOLUME II

N. Ackerman
[2012] *On n-cardinal spectra of ultrahomogeneous theories*, preprint.

R. Akhtar and A. H. Lachlan
[1995] *On countable homogeneous 3-hypergraphs*, **Archive for Mathematical Logic**, vol. 34, pp. 331–344.

D. Amato, G. Cherlin, and H. D. Macpherson
[2021] *Metrically homogeneous graphs of diameter 3*, **Journal of Mathematical Logic**, vol. 21, no. 1, pp. 1–106, paper No. 2050020.

A. Aranda, D. Bradley-Williams, E. K. Hng, J. Hubička, M. Karamanlis, M. Kompatscher, M. Konečný, and M. Pawliuk
[2021] *Completing graphs to metric spaces*, **Contributions to Discrete Mathematics**, vol. 16, pp. 71–89, e-print arXiv:1706.00295 [math.CO], 2017.

A. Aranda, D. Bradley-Williams, J. Hubička, M. Karamanlis, M. Kompatscher, M. Konečný, and M. Pawliuk
[2017] *Ramsey expansions of metrically homogeneous graphs*, e-print arXiv:1707.02612 [math.CO].

P. Baginski
[2009] **Stable $\aleph_0$-Categorical Algebraic Structures**, Ph.D. thesis, University of California, Berkeley, Ann Arbor, MI, 132 pp.

W. Baur, G. Cherlin, and A. Macintyre
[1979] *Totally categorical groups and rings*, **Journal of Algebra**, vol. 57, pp. 407–440.

I. Ben Yaacov, J. Melleray, and T. Tsankov
[2017] *Metrizable universal minimal flows of Polish groups have a comeagre orbit*, **Geometric and Functional Analysis**, vol. 27, no. 1, pp. 67–77.

A. Bishnoi and B. De Bruyn
[2016] *On semi-finite hexagons of order $(2, t)$ containing a subhexagon*, **Annals of Combinatorics**, vol. 20, no. 3, pp. 433–452.

M. BODIRSKY

[2015] *The complexity of constraint satisfaction problems*, **32nd International Symposium on Theoretical Aspects of Computer Science** (Wadern), LIPIcs: Leibniz International Proceedings in Informatics, Schloss Dagstuhl, vol. 30, Leibniz-Zentrum für Informatik, pp. 2–9.

[2021] **Complexity of Infinite-Domain Constraint Satisfaction**, Lecture Notes in Logic, vol. 52, Cambridge University Press, Cambridge; Association for Symbolic Logic, Ithaca, NY.

M. BODIRSKY, S. KNÄUER, AND J. RYDVAL

[2021] *Amalgamation is PSPACE-hard*, e-print arXiv:2108.00452 [cs.LO]; revised.

M. BODIRSKY AND M. PINSKER

[2011] *Reducts of Ramsey structures*, **Model Theoretic Methods in Finite Combinatorics** (M. Grohe and J. A. Makowsky, editors), Contemporary Mathematics, no. 558, American Mathematical Society, Providence, RI, pp. 489–519.

M. BODIRSKY, M. PINSKER, AND A. PONGRÁCZ

[2015] *The 42 reducts of the random ordered graph*, **Proceedings of the London Mathematical Society. Third Series**, vol. 111, no. 3, pp. 591–632.

M. BODIRSKY, M. PINSKER, AND T. TSANKOV

[2013] *Decidability of definability*, **The Journal of Symbolic Logic**, vol. 78, no. 4, pp. 1036–1054.

M. BODIRSKY, J. RYDVAL, AND A. SCHROTTENLOHER

[2021] *Universal Horn sentences and the joint embedding property*, e-print arXiv:2104.11123 [cs.LO].

A. BONATO AND D. DELIĆ

[1997/98] *The model companion of width-two orders*, **Order**, vol. 14, pp. 87–99.

S. BRAUNFELD

[2016] *The lattice of definable equivalence relations in homogeneous n-dimensional permutation structures*, **Electronic Journal of Combinatorics**, vol. 23, no. 4, paper 44, 24 pp.

[2019] *The undecidability of joint embedding and joint homomorphism for hereditary graph classes*, **Discrete Mathematics & Theoretical Computer Science**, vol. 21, no. 2, paper 9, 17 pp.

P. J. CAMERON

[1990] **Oligomorphic Permutation Groups**, London Mathematical Society Lecture Note Series 152, Cambridge University Press.

[1998] *A census of infinite distance transitive graphs*, **Discrete Mathematics**, vol. 192, pp. 11–26.

[2000] *Homogeneous Cayley objects*, **European Journal of Combinatorics**, vol. 21, pp. 745–760, Special volume: Discrete metric spaces (Marseille, 1998).

[2002/03] *Homogeneous permutations*, **Electronic Journal of Combinatorics**, vol. 9, paper 2, pp. 1–9.

P. J. CAMERON AND K. W. JOHNSON
[1987] *An investigation of countable B-groups*, **Mathematical Proceedings of the Cambridge Philosophical Society**, vol. 102, pp. 223–231.

P. J. CAMERON AND A. M. VERSHIK
[2006] *Some isometry groups of the Urysohn space*, **Annals of Pure and Applied Logic**, vol. 143, pp. 70–78.

G. CHERLIN
[1987] *Homogeneous directed graphs I. The imprimitive case*, **Logic Colloquium 1985** (The Paris Logic Group, editor), North-Holland, Amsterdam, pp. 67–88.

[1998] *The Classification of Countable Homogeneous Directed Graphs and Countable Homogeneous n-Tournaments*, **Memoirs of the American Mathematical Society**, vol. 131, no. 621, xiv+161 pp.

[1999] *Infinite imprimitive homogeneous 3-edge-colored complete graphs*, **The Journal of Symbolic Logic**, vol. 64, pp. 159–179.

[2000] *Sporadic homogeneous structures*, **The Gelfand Mathematical Seminars, 1996–1999**, Gelfand Math. Sem., Birkhäuser Boston, Boston, MA, pp. 15–48.

[2005] *Locally finite generalized quadrangles with at most five points per line*, **Discrete Mathematics**, vol. 291, pp. 73–79.

[2011a] *Forbidden substructures and combinatorial dichotomies: WQO and universality*, **Discrete Mathematics**, vol. 311, pp. 1543–1584.

[2011b] *Two problems on homogeneous structures, revisited*, **Model Theoretic Methods in Finite Combinatorics** (M. Grohe and J. A. Makowsky, editors), Contemporary Mathematics, no. 558, American Mathematical Society, Providence, RI, pp. 319–415.

[2016] *On the relational complexity of a finite permutation group*, **Journal of Algebraic Combinatorics**, vol. 43, pp. 339–374.

[2021] *Homogeneity and related topics: An extended bibliography*, e-print arXiv:2111.15429 [math.LO].

G. CHERLIN AND E. HRUSHOVSKI
[2003] **Finite Structures with Few Types**, Annals of Mathematics Studies, vol. 152, Princeton University Press, Princeton, NJ, vi+193 pp.

G. CHERLIN, J. HUBIČKA, M. KONEČNÝ, AND J. NEŠETŘIL
[2021] *Ramsey expansions of 3-hypertournaments*, **Extended Abstracts EuroComb 2021** (Jaroslav Nešetřil, Guillem Perarnau, Juanjo Rué, and Oriol

Serra, editors), European Conference on Combinatorics, Graph Theory and Applications, Springer, pp. 696–701.

G. CHERLIN, S. SHELAH, AND N. SHI
[1999] *Universal graphs with forbidden subgraphs and algebraic closure*, **Advances in Applied Mathematics**, vol. 22, pp. 454–491.

G. CHERLIN AND L. TALLGREN
[2007] *Universal graphs with a forbidden near-path or 2-bouquet*, **Journal of Graph Theory**, vol. 56, pp. 41–63.

G. CHERLIN AND S. THOMAS
[2002] *Two cardinal properties of homogeneous graphs*, **The Journal of Symbolic Logic**, vol. 67, pp. 217–220.

G. CONANT
[2015] **Model Theory and Combinatorics of Homogeneous Metric Spaces**, Ph.D. thesis, University of Illinois at Chicago, 288 pp.
[2017] *Distance structures for generalized metric spaces*, **Annals of Pure and Applied Logic**, vol. 168, no. 3, pp. 622–650.
[2019] *Extending partial isometries of generalized metric spaces*, **Fundamenta Mathematicae**, vol. 244, no. 1, pp. 1–16.

R. COULSON
[2019] **Metrically Homogeneous Graphs: Dynamical Properties of their Automorphism Groups and the Classification of Twists**, Ph.D. thesis, Rutgers University.

J. COVINGTON
[1990] *Homogenizable relational structures*, **Illinois Journal of Mathematics**, vol. 34, pp. 731–743.

F. DALLA VOLTA, N. GILL, AND P. SPIGA
[2018] *Cherlin's conjecture for sporadic simple groups*, **Pacific Journal of Mathematics**, vol. 297, no. 1, pp. 47–66.

I. DOLINKA AND D. MAŠULOVIĆ
[2012] *Countable homogeneous linearly ordered posets*, **European Journal of Combinatorics**, vol. 33, pp. 1965–1973.

M. EL-ZAHAR AND N. SAUER
[1991] *Ramsey-type properties of relational structures*, **Discrete Mathematics**, vol. 94, pp. 1–10.
[1993] *On the divisibility of homogeneous directed graphs*, **Canadian Journal of Mathematics**, vol. 45, pp. 284–294.

D. EVANS, J. HUBIČKA, M. KONEČNÝ, AND J. NEŠETŘIL
[2020] *EPPA for two-graphs and antipodal metric spaces*, **Proceedings of the American Mathematical Society**, vol. 148, no. 5, pp. 1901–1915.

D. M. EVANS, J. HUBIČKA, M. KONEČNÝ, AND Y. LI
[2019] *Simplicity of the automorphism groups of generalised metric spaces*,
e-print arXiv:1907.13204 [math.GR].

D. M. EVANS, J. HUBIČKA, AND J. NEŠETŘIL
[2019] *Automorphism groups and Ramsey properties of sparse graphs*,
Proceedings of the London Mathematical Society. Third Series, vol. 119, no. 2,
pp. 515–546.

CH. EVEN-ZOHAR AND N. LINIAL
[2015] *Triply existentially complete triangle-free graphs*, **Journal of Graph Theory**, vol. 78, pp. 305–317.

U. FELGNER
[1978] $\aleph_0$-*categorical stable groups*, **Mathematische Zeitschrift**, vol. 160, pp. 27–49.

N. GILL, F. HUNT, AND P. SPIGA
[2017] *Cherlin's conjecture for almost simple groups of Lie rank* 1, e-print
arXiv:1705.01344 [math.GR].

N. GILL, M. LIEBECK, AND P. SPIGA
[2021] *Cherlin's conjecture for finite groups of Lie type*, e-print
arXiv:2106.05154 [math.GR].

N. GILL AND P. SPIGA
[2016] *Binary permutation groups*: *Alternating and classical groups*, e-print
arXiv:1610.01792 [math.GR].

M. GOLDSTERN, R. GROSSBERG, AND M. KOJMAN
[1996] *Infinite homogeneous bipartite graphs with unequal sides*, **Discrete Mathematics**, vol. 149, pp. 69–82.

C. WARD HENSON
[1971] *A family of countable homogeneous graphs*, **Pacific Journal of Mathematics**, vol. 38, pp. 69–83.

E. HRUSHOVSKI
[2020] *Definability patterns and their symmetries*, Preprint.

J. HUBIČKA
[2019] *Structural Ramsey theory and the extension property for partial automorphisms*, Introduction to habilitation, 35 pp.

J. HUBIČKA, M. KONEČNÝ, AND J. NEŠETŘIL
[2020a] *Semigroup-valued metric spaces*: *Ramsey expansions and EPPA*, in
preparation.
[2020b] *All those EPPA classes* (*Strengthenings of the Herwig-Lascar theorem*), e-print, arXiv:1902.03855 [math.CO]. v2, 2020.

J. HUBIČKA AND J. NEŠETŘIL
[2019] *All those Ramsey classes (Ramsey classes with closures and forbidden homomorphisms)*, **Advances in Mathematics**, vol. 356, pp. 106791, 89.

M. JUNKER AND M. ZIEGLER
[2008] *The 116 reducts of* $(\mathbb{Q}, <, a)$, **The Journal of Symbolic Logic**, vol. 73, pp. 861–884.

W. KANTOR
[1972] *k-Homogeneous groups*, **Mathematische Zeitschrift**, vol. 124, pp. 261–265.

A. KECHRIS, V. PESTOV, AND S. TODORCEVIC
[2005] *Fraïssé limits, Ramsey theory, and topological dynamics of automorphism groups*, **Geometric and Functional Analysis**, vol. 15, pp. 106–189.

A. KECHRIS AND C. ROSENDAL
[2007] *Turbulence, amalgamation, and generic automorphisms of homogeneous structures*, **Proceedings of the London Mathematical Society (3)**, vol. 94, pp. 302–350.

P. KOMJÁTH AND J. PACH
[1991] *Universal elements and the complexity of certain classes of infinite graphs*, **Discrete Mathematics**, vol. 95, pp. 255–270.

M. KONEČNÝ
[2019a] *Combinatorial properties of metrically homogeneous graphs*, Bachelor's thesis, Charles University, Prague.
[2019b] *Semigroup-valued metric spaces*, Master's thesis, Charles University, Prague.
[2020] *Extending partial isometries of antipodal graphs*, **Discrete Mathematics**, vol. 343, no. 1, pp. 111633, 13.

V. KOPONEN
[2016] *Binary simple homogeneous structures are supersimple with finite rank*, **Proceedings of the American Mathematical Society**, vol. 144, no. 4, pp. 1745–1759.
[2017] *Binary primitive homogeneous simple structures*, **The Journal of Symbolic Logic**, vol. 82, no. 1, pp. 183–207.

A. H. LACHLAN
[1984] *Countable homogeneous tournaments*, **Transactions of the American Mathematical Society**, vol. 284, pp. 431–461.
[1986] *Binary homogeneous structures II*, **Proceedings of the London Mathematical Society**, vol. 52, pp. 412–426.
[ca. 1982] *Verification of the classification of stable homogeneous 3-graphs*, undated preprint.

Y. Li
[2019] *Automorphism groups of homogeneous structures with stationary weak independence relations*, e-print arXiv:1911.08540 [math.GR].

J. Linman and M. Pinsker
[2015] *Permutations on the random permutation*, **The Electronic Journal of Combinatorics**, vol. 22, no. 2, paper 54, 22 pp.

H. D. Macpherson
[1986] *Homogeneity in infinite permutation groups*, **Periodica Mathematica Hungarica**, vol. 17, pp. 211–233.
[1991] *Finite axiomatizability and theories with trivial algebraic closure*, **Notre Dame Journal of Formal Logic**, vol. 32, pp. 188–192.
[2011] *A survey of homogeneous structures*, **Discrete Mathematics**, vol. 131, pp. 1599–1634.

H. D. Macpherson and K. Tent
[2011] *Simplicity of some automorphism groups*, **Journal of Algebra**, vol. 342, pp. 40–52.

V. D. Mazurov and E. I. Khukhro
[2014] *Unsolved problems in group theory. the Kourovka notebook. no. 18*, e-print, arXiv:1401.0300 [math.GR].

J. Melleray, L. Nguyen Van Thé, and T. Tsankov
[2016] *Polish groups with metrizable universal minimal flows*, **International Mathematics Research Notices. IMRN**, no. 5, pp. 1285–1307.

L. Moss
[1992] *Distanced graphs*, **Discrete Mathematics**, vol. 102, pp. 287–305.

J. Nešetřil
[1979] *Amalgamation of graphs and its applications*, **Second International Conference on Combinatorial Mathematics (New York, 1978)** (New York), Annals of New York Academy of Sciences, vol. 319, New York Academy of Sciences, pp. 415–428.

D. Palacín
[2017] *Generalized amalgamation and homogeneity*, **The Journal of Symbolic Logic**, vol. 82, no. 4, pp. 1409–1421.

M. Rubin
[1994] *On the reconstruction of $\aleph_0$-categorical structures from their automorphism groups*, **Proceedings of the London Mathematical Society**, vol. 69, pp. 225–249.

N. Ruškuc
[2005] *Decidabilty questions for pattern avoidance classes of permutations*, Talk at Gainesville (slides).

D. SARACINO
  [1999] *On a combinatorial problem from the model theory of wreath products.*
I, II, Journal of Combinatorial Theory Series A, vol. 86, pp. 281–305, 306–322.
  [2000] *On a combinatorial problem from the model theory of wreath products.*
III, Journal of Combinatorial Theory Series A, vol. 89, pp. 231–269.

P. SIMON
  [2021] *NIP ω-categorical structures: the rank 1 case*, e-print arXiv:180
7.07102 [math.LO].

T. SKOLEM
  [1920] *Logisch-kombinatorische Untersuchungen über die Erfüllbarkeit und
Beweisbarkeit mathematischer Sätze nebst einem Theoreme über dichte Mengen,
Videnskapsselskapets Skrifter I. Matematisk-naturvidenskapelig Klasse*, vol. 4,
pp. 1–36, see §4: "Ein Satz über dichte Mengen".

K. TENT AND M. ZIEGLER
  [2013a] *The isometry group of the bounded Urysohn space is simple, Bulletin
of the London Mathematical Society*, vol. 45, no. 5, pp. 1026–1030.
  [2013b] *On the isometry group of the Urysohn space, Journal of the London
Mathematical Society. Second Series*, vol. 87, no. 1, pp. 289–303.

S. THOMAS
  [1991] *Reducts of the random graph, The Journal of Symbolic Logic*, vol. 56,
pp. 176–181.
  [1998] *The nonexistence of a binary homogeneous pseudoplane, Mathematical
Logic Quarterly*, vol. 44, pp. 135–137.

J. WISCONS
  [2016] *A reduction theorem for primitive binary permutation groups, Bulletin
of the London Mathematical Society*, vol. 48, pp. 291–299.

INDEX

Boldface page numbers refer to definitions.